SCIENCE AND
THE NEW CIVILIZATION

SCIENCE AND
THE NEW CIVILIZATION

BY

ROBERT A. MILLIKAN

Essay Index Reprint Series

 BOOKS FOR LIBRARIES PRESS
FREEPORT, NEW YORK

INTERNATIONAL STANDARD BOOK NUMBER:
0-8369-2418-5

LIBRARY OF CONGRESS CATALOG CARD NUMBER:
76-142671

PRINTED IN THE UNITED STATES OF AMERICA
BY
NEW WORLD BOOK MANUFACTURING CO., INC.
HALLANDALE, FLORIDA 33009

INTRODUCTORY NOTE

This book consists of a series of addresses
some of which have appeared before in the
form of articles, but all of which represent discussion of one aspect or another of the central
theme chosen as the title of the book. Acknowl-
edgments are due to *The Atlantic Monthly*, to
The Forum, to *Science*, and to *Scribner's Maga-
zine* for permission to reproduce in the present
form discussions which have appeared in part
in their pages.

Though the speaker is a scientist, he is ad-
dressing herein a general rather than a scientific
audience in the hope that the presentation in
this form of his point of view may assist in the
spread of a better understanding of the purposes and attitude of science, as well as of the
rôle which it has played in the past, is playing
now, and may yet play in the progress of the
race.

CONTENTS

SCIENCE AND
THE NEW CIVILIZATION

I

SCIENCE AND MODERN LIFE

In the summer of 1927 it was my lot to be called out of my laboratory to attend in rapid succession (1) a meeting of the Committee on Intellectual Co-operation of the League of Nations at Geneva, a body called into being for the sake of assisting in laying better foundations for international good-will and understanding than have heretofore existed; (2) the annual meeting of the British Association for the Advancement of Science at Leeds, one of the most important of the Old World's scientific bodies, whose meetings have marked the milestones of scientific progress; and (3) the International Congress of Physics held at Como and at Rome in commemoration of the hundredth anniversary of the death of Alessandro Volta, the discoverer of current electricity, and thus, in a certain sense, the initiator of this amazing electrical century—suitable errands to inspire reflections on the place of science in modern life. I should

[1]

like to present them in the form of a few pictures.

As we sped, a thousand persons across the Atlantic in an oil-burning ship in which even the modern stoker—whose "hard fate" has often been held up as a symbol of the evils of our "mechanical age"—has now a comfortable and an interesting job, for he simply and quietly guides the expenditure of hundreds of thousands of man-power represented in the energy of separated hydrogen and oxygen and carbon atoms rushing eagerly together to fulfill their predetermined destiny, and merely incidentally in so doing sending the ship racing across the Atlantic—in the face of that situation, could I, or could any one not completely blind to the significance of modern life, fail to reflect somewhat as follows? If Cicero, or Pericles, or any man of any preceding civilization, had been sent on a similar errand, had he had any power at all except the winds, it would have been the man-power furnished to the triremes by the straining sinews of hundreds of human slaves chained to their oars, slaves to be simply cast aside into the sea, if they weakened or gave out, and then replaced by other slaves! Could any man fail to reflect that our scientific civilization is the first one in history which has not been built on just such human slavery, the first which offers the hope, at least, and a hope already partially realized, of relieving mankind forever from the worst of the physical bondage with which all civilizations have heretofore enchained him, whether it be the slavery represented by Millet's man with the hoe—a dumb beast-like broken-backed agricultural drudge—or the slavery at the galley pictured in "Ben Hur," or the slavery of the pyramid builders referred to in the books of Moses?

Or again, could any one who stood with me at the base of the column of Trajan, matchless relic and symbol of the unequalled magnificence of Rome, fail to muse first that ancient man in the immensity and daring of his undertakings, in the grandeur of his conceptions, in the beauty and skill of his workmanship, in his whole intellectual equipment was fully our equal if not our superior?—for we shall leave no monuments like his. But could he also fail to reflect that ancient man built these monuments solely through the unlimited control of enforced human labor, while we have not only freed that slave but have made him the master and director of the giant but insensible Titans of the lower world? We call them now by the unromantic names Coal and Oil. It is our triumph over these that have given to *him* freedom and opportunity. This is one side of the picture of science and the modern world, a side that can be presented with a thousand variants but all having the same inspiring significance.

Another picture. In a comfortable English home out in the country in north England a small group is seated, sipping after-dinner coffee, enjoying conversation, and interrupting it now and then to listen to something particularly fine that is coming in over the radio. The technic of the reproduction is superb, but no more so than that with which we are familiar in our American homes, for the whole broadcasting idea, as well as the main part of its technical development, is American in its origins. But the programme that is on the air in England is incomparably superior to anything to be heard here, for the English government has taken

[4]

over completely the control of the radio. It col-
lects from each owner of a receiving set twelve
shillings a year, and then, with the large funds
thus obtained,—for there are as many radio
fans in England as in America,—it provides the
radioland-public of England with the largest
return in education and in entertainment for
eight mills a night ever provided, I suspect,
anywhere in the history of the world. For it
employs only high-class speakers, musicians, and
entertainers of all sorts, so that the whole British nation is now being given educational advan-
tages of the finest possible sort through the radio, at less than a cent a family a night, collected
only from those who wish to take advantage of
them.

Nor is it merely the subject matter of the radio programmes that is commendable. The value of giving the whole British public the opportunity to hear the English language used, in in-
tonations and otherwise, as cultured people are
wont to use it, is altogether inestimable. And
sitting there in the north of England we had but
to turn the dial to the wavelength used by Berlin and we heard an equally authoritative use of

the German language, and I envisaged a whole population, or as many of them as wished, learning a new language, easily and correctly, instead of through the stupidities of grammar, as we now go at it. What a stimulant to the imagination! What possibilities are here, only just beginning to be realized, for public education, for the enrichment of the life of the country dweller, as well as the city resident, solely because of such an influence as this of *modern physics upon modern life.*

Now turn to another picture which presents the other side of the story. Sir Arthur Keith, the foremost British anthropologist, was then the president of the British Association. The Leeds meeting represented the fiftieth anniversary of the meeting at which Darwin's then new theory of evolution had been first vigorously debated. Sir Arthur took at this session as the subject of his presidential address, "Darwin's Theory of Man's Descent as It Stands To-day." He showed that fifty years of fossil study had given extraordinary confirmation to the general outline of the evolutionary conception, had placed it indeed upon well-nigh impregnable foundations.

[6]

The following Sunday the Bishop of Ripon preached upon science and modern life. He thought we were gaining new scientific knowledge, and acquiring control of stupendous new forces, faster than we were developing our abilities to control ourselves, faster than we were exhibiting capacity to be intrusted with these new forces, and hence he suggested that science as a whole take a ten-year holiday.

When, the next day, the newspaper men, who had had as good a story out of the whole incident as our newspapers got out of the Scopes trial, pressed the Bishop to define more sharply just what he meant by a ten-year scientific holiday, he said that he thought the workers in medicine and in public health ought not to stop, since then the germs of disease might steal a march on us, and avoidable suffering be thereby caused. He had had in mind, rather, a vacation for physics and chemistry and the parts of biology not associated with the improvement of health and the alleviation of suffering.

The Bishop's explanation is of value as throwing an illuminating side light upon the sort of emotionalism and misunderstanding that is represented in much of the present public antagonism to scientific progress. The question which the Bishop raises is proper enough, but the conclusion is altogether incorrect. For, first, physics and chemistry cannot take a holiday without turning off the power on all the other sciences that depend upon them; and, second, physics, chemistry, and genetics are, in fact, the great constructive sciences which alone stand between mankind and its dire fate foreseen by Malthus; while without these the palliative sciences, such as the Bishop mentioned, only hasten and make inevitable the horrors of that day.

The incident is presented because it is illustrative of a wide-spread attitude as to the danger of flooding the world with too much knowledge. The fear of knowledge is quite as old as the Garden of Eden. Prometheus was chained to a rock and had his liver torn out by a vulture because he had dared to steal knowledge from the Gods and bring it down to men. The story of Faust, which permeates literature up to within a hundred years, is evidence of the wide-spread, age-long belief in the liaison between the man of knowledge and the powers of darkness. It will persist so long as superstition, as distinct from reverence, lasts.

But there is a real question, not to be thus easily disposed of, which the Bishop's sermon puts before the man of science. It is this. Am I myself a broadly enough educated man to distinguish, when I am engaged in the work of reconstruction, between the truth of the past and the error of the past, and not to pull them both down together? Am I sufficiently familiar with what the past has learned, and what it therefore actually has to teach, and am I enough of a *statesman* not to remove any brick from the structure of man's progress until I see how to replace it by a better one? I am sorry to be obliged to admit that some of us scientists will have to answer that question in the negative. Such justification as there may be for the public's distrust of science is due chiefly to the misrepresentation of science by some of its *uneducated* devotees. For men without any real understanding are of course to be found in all the walks of life.

This problem, however, is not at all peculiar to science. In fact, the most wantonly destructive forces in modern life, and the most sordidly commercial, are not in general found in the field of science nor having anything to do with it. It is literature and art, much more than science, which have been the prey of those influences through which the chief menace to our civilization comes. After the law of gravitation, or the principle of conservation of energy, have been once discovered and established, physics understands quite well that *its future progress must be made in conformity with these laws,* at least that Einstein must include Newton, and it succeeds fairly well in keeping its levitators and its inventors of devices for realizing perpetual motion under suitable detention, or restraint, somewhere. But society has as yet developed no protection against its perpetual motion cranks (the devotees of the new, regardless of the true) in the fields of literature and art, and that despite the fact that sculpture has had its Phidias and literature its Shakespeare just as truly as physics has had its Newton or biology its Pasteur. I grant that in literature and art, and in non-scientific fields generally, it is more difficult than in science to know what has been found to be truth and what error, that in many cases we don't yet know; yet there are even here certain broad lines of established truth recognized by thoughtful people everywhere. For example, the race long ago learned that unbridled license in the individual is incompatible with social progress, that civilization, which is orderly group life, will perish and the race go back to the jungle unless the sense of social responsibility can be kept universally alive. And yet to-day literature is infested here and there with unbridled license, with emotional, destructive, over-sexed, neurotic influences, the product of men who are either incompetent to think anything through to its consequences, or else who belong to that not inconsiderable group who protest that they are not in the least interested in social consequences anyway, men who, in their own words, are merely desirous of "expressing themselves." Such men are, in fact, nothing but the perpetual motion cranks of literature and of art. It is from this direction, not from the direction of science that the chief menaces to our civilization are now coming.

But despite this situation I should hesitate to

suggest that all writers and all artists be given a holiday. This is an age of specialization, and properly so, and some evils from our specialization are to be expected. Our job is to minimize them and to find counterirritants for them. I am not altogether discouraged even when I find a humanist of the better sort who is only half educated. Let this incident illustrate. Not long ago I heard a certain literary man of magnificent craftsmanship and fine influence in his own field, declare that he saw no values in our modern "mechanical age." Further, this same man recently visited a plant where the very foundations of our modern civilization are being laid. A ton of earth lies underneath a mountain. Scattered through that ton in infinitesimal grains is just two dollars' worth of copper. That ton of earth is being dug out of its resting place, transported to the mill miles away, the infinitesimal particles of copper miraculously picked out by invisible chemical forces, then deposited in great sheets by the equally invisible physical forces of the electric current, then shipped three thousand miles and again refined, then drawn into wires to transport the formerly wasted energy of a waterfall, and all these operations from the buried ton of Arizona dirt to refined copper in New York done at a cost of less than two dollars, for there was no more value there.

This amazing achievement not only did not interest the humanist, but he complained about disfiguring the desert by electrical transmission lines. Unbelievable blindness—a soul without a spark of imagination else it would have seen the hundred thousand powerful, prancing horses which are speeding along each of those wires, transforming the desert into a garden, making it possible for him and his kind to live and work without standing on the bowed backs of human slaves as his prototype has always done in ages past. Seen in this rôle, that humanist was neither humanist nor philosopher, for he was not really interested in humanity. In this picture it is the scientist who is the real humanist. Nevertheless, the Bishop of Ripon was right enough in distrusting the wisdom, and sometimes even the morality, of individual scientists, and of individual humanists, too. But the remedy is certainly not "to give science a holiday." That is both impossible and foolish. It is rather *to*

*reconstruct and extend our educational processes
so as to make broader-gauge and better educated
scientists and humanists alike.* There is no other
remedy.

But, says some one, these pictures so far deal
only with the superficial aspects of life. What
has science to say to him whose soul is hungry,
to him who cries "Man shall not live by bread
alone"? Has it anything more than a dry crust
to offer him? The response is instant and unambiguous. Within the past half century, as
a direct result of the findings of modern science,
there has developed an evolutionary philosophy
—an evolutionary religion, too, if you will—
which has given a new emotional basis to life,
the most inspiring and the most forward-looking that the world has thus far seen. For, first,
the findings of physics, chemistry, and astronomy have within twenty-five years brought to
light a universe of extraordinary and unex-
pected orderliness, and of the wondrous beauty
and harmony that go with order. It is the same
story whether one looks out upon the island
universes brought to light by modern astronomy,
and located definitely, some of them, a million

light years away, or whether he looks down into the molecular world of chemistry, or through it to the electronic world of physics, or peers even inside the unbelievably small nucleus of the atom. Also, in the organic world, the sciences of geology, paleontology, and biology have revealed, still more wonderfully, an orderly development from lower up to higher forms, from smaller up to larger capacities—a development which can definitely be seen to have been going on for millions upon millions of years and which therefore gives promise of going on for ages yet to be.

> "A fire mist and a planet,
> A crystal and a cell,
> A jelly fish and a saurian,
> And caves where cavemen dwell.
> Then a sense of law and beauty,
> And a face turned from the clod.
> Some call it evolution
> And others call it God."

That sort of sentiment is the gift of modern science to the world.

And there is one further finding of modern science which has a tremendous inspirational appeal. It is the discovery of the vital part which we ourselves are playing in this evolutionary process. For man himself has within two hundred years discovered new forces with the aid of which he is now consciously and very rapidly making over both his physical and his biological environment. The Volta Centenary, a symbol of our electrical age, was representative of the one, the stamping out of yellow fever is an illustration of the other. And if the biologist is right that the biological evolution of the human organism is going on so slowly that man himself is not now endowed with capacities appreciably different from those which he brought with him into the period of recorded history, then, since, within this period the forward strides that he has made in his control over his environment, in the development of his civilization, have been stupendous and unquestionable, it follows that this progress has been due, not to the betterment of his stock, but rather primarily *to the passing on of the accumulated knowledge of the race to the generations following after. The great instruments of progress for mankind are then research, the discovery of*

*new knowledge, and education, the passing on
of the store of accumulated wisdom to our followers.* This puts the immediate destinies of the
race, or of our section of the race, or of our section of our country, largely in our own hands.
This spirit and this conviction is the gift of
modern science to the world. Is it then too
much to say that modern science has remade
philosophy and revivified religion?

The next picture brings into the foreground
what I regard as the most important contribution of science to modern life. The scene is laid
in Geneva, the occasion, a meeting of the Council of the League of Nations. The speaker is
Nansen, the tall, white haired, rugged faced,
heavily mustached Norwegian explorer, now
directing the tamed and controlled energies of
his fierce, Viking blood to trying to find a solution to the tragic situation of the Armenians, a
situation to which heretofore there has been no
solution except extermination. After four years
of effort he brings in a discouraging report, and
thinks the League of Nations must write down
"the record of its first failure." He requests
the Council to strike the Armenian matter from

[17]

its programme, promising, however, to keep at it himself and to try through other agencies to find a solution. Then Briand of France speaks. Quietly he begs Nansen not yet to despair of the League's assistance. He is sure some solution can be found, and promises that his country, in financial straits though it be, will not be lacking in lending its assistance. The representatives of other nations follow in similar vein, the problem is retained on the Council's programme, and the conviction is at least fortified that *with the right kind of attack a solution may yet be found to an age-old difficulty—that extermination is not the only answer to race rivalry.*

With the right sort of attack! What is it that the League of Nations as a whole is trying to do? It is trying for the first time in human history to use *the objective mode of approach* to international difficulties in the conviction that there is some better solution than the arbitrament of war. But whence has come that conviction? It is itself a contribution of modern science to modern life. In the days of the jungle war *was* the best solution,—at least it was the only solution. It was Nature's way of enabling the fittest to survive, and we are not so far past the days of the jungle yet. Within fifty years as great an historian as Edward Meyer, and as great a humanist as John Ruskin, have lauded war as the finest developer of a people. *But the growth of modern science has demonstrated that war is no longer, in general, the best way to enable the fittest to survive.* The great war profited no one. It injured all the main participants. Modern science has created a new world in which the old rules no longer work.

> "New occasions teach new duties,
> Time makes ancient good uncouth,
> He must upward still and onward
> Who would keep abreast of truth."

Alfred Nobel was perhaps not far from right when he thought that he had taken the main step to the abolition of war by the invention of nitroglycerine. He has, I suspect, exerted a larger influence in that direction than have all the sentimental pacificist organizations that have ever existed. For sentimental pacificism is, after all, but a return to the method of the jungle.

[19]

It is in the jungle that emotionalism alone determines conduct, and *wherever that is true no other than the law of the jungle is possible.* For the emotion of hate is sure sooner or later to follow the emotion of love, and then there is a spring for the throat. It is altogether obvious that the only quality which really distinguishes man from the brutes is his reason. You may call that an unsafe guide, *but he has absolutely no other* unless he is to turn his face back toward the jungle.

The reason that war must be largely, if not wholly, eliminated is that, because of the growth of modern science, it is no longer safe to allow a country to engage in aggressive war. The risks to its own life and to the life of the race are too great, as the last war showed. The nations of the world *are forced by the advances in modern science to find some other way* to solve their international difficulties. Sentimental pacifism has nothing to do with it. It is a hindrance rather than a help. It acts rather as a stimulant than as a deterrent to war by rendering the dangers to the aggressor less real than they otherwise would be. There is no sort of alternative except to set up in international matters precisely as we have already done in intercommunity and interstate affairs some sort of an organization for making studies by the objective method of international difficulties and finding other solutions.

But what exactly do I mean by the objective method? Somebody has said that "What we call the process of reasoning is merely the process of rearranging one's prejudices," and we admit the truth of this assertion when we say, as we so often do, "Oh, yes, I understand that is the excuse, but what is, after all, the reason?" Indeed, there is no question but that a large part of what we call reasoning is in fact simply the rearranging of prejudices. In so far, for example, as we are Republicans, or Democrats, or Presbyterians, or Catholics, or Mohammedans, or prohibitionists, because our fathers bore those brands, and many of us will be admitted *by our acquaintances*, at least, to have no other real grounds for our labels, our so-called reasonings on these subjects certainly consist in nothing more than the rearrangement of our prejudices. The lawyer who takes a case first, and develops his argument later, is obviously only rearranging his preconceptions.

If, however, one wishes to obtain a clear idea of what the objective method is, he has only to become acquainted with the way in which all problems are attacked in the analytical sciences. In physics, for example, the procedure in problem solving is always first to collect the facts, *i. e.*, to make the observations with complete honesty and complete disregard of all theories and all presuppositions, and then to analyze the data to see what conclusions follow necessarily from them, or what interpretations are consistent with them. This method, while not confined at all to the physical sciences, is nevertheless commonly known as the scientific method in recognition of the fact that it has had its fullest development and its most conspicuous use in the sciences. Indeed, I regard the development and spread of this method as the most important contribution of science to life, for *it represents the only hope of the race of ultimately getting out of the jungle*. The method can in no way be acquired and understood so well as by the study of the analytical sciences, and hence an education which has left out these sciences has, in my judgment, lost the most vital element in all education. Nor is that merely the individual judgment of a prejudiced scientist, as the following quotation from William B. Munro, one of our most prominent humanists, a member of the faculty of the Harvard University, shows:

"It is the glory of pure science and of mathematics that these subjects train men in orderly and objective thinking as no other subjects can. Here are fields of study in which loose or crooked thought leads inevitably to demonstrable error, to error which cannot be glossed over or concealed. Here are branches of knowledge in which there is no confusion between right and wrong, between post hocs and propter hocs, between the mere coincidences and the consequences of a cause. When you have finished with a problem in any of the exact sciences you are either right or wrong, and you know it. That is why we call them exact sciences, to distinguish them from philosophy, sociology, economics, and the other social sciences, in which the difference between truth and

[23]

error is still, in most cases, a matter of individual opinion. Many years ago physics was known as 'natural philosophy'; it was merely a body of speculative ideas concerning the mechanics of nature. It became an exact science by developing an inductive methodology, which makes all the difference between science and guesswork.

"Some years ago, in the Harvard Law School, we thought it worth while to inquire into the educational antecedents of the student body, with a view to ascertaining whether there was any relation between success in the study of law and the previous collegiate training of these young men. In the Harvard Law School there are more than a thousand students, all of them college graduates, drawn from every section of the country. Nearly all of them have specialized, during their undergraduate years, in some single subject or group of subjects—languages, history, science, philosophy, economics, mathematics, and so on. Offhand one would probably say that the young man who had devoted most of his attention to the study of history, government, and economics while in college would be gaining the best preparation for the study of law—for these are the subjects which in their content come nearest to the law; but that is not what we found. On the contrary the results of this inquiry showed that the young men who had specialized in ancient languages, in the exact sciences, and especially in mathematics, were on the whole better equipped for the study of law, and were making higher rank in it, than were those who had devoted their energies to subjects more closely akin."

But can education, even in the sciences, do the work fast enough to prevent the catastrophe feared by the Bishop of Ripon? Can we learn to control our emotions and impulses and our new-found powers, to take the long view, and to do the rational thing instead of the emotional, or the vicious, thing with the enormous forces given to us by science? Can we alter human nature?

Perhaps the following is a partial answer: Twenty-five years ago if any one had asked you or me or any body of men however intelligent whether human nature could be so altered in a reasonable time as to make it safe to intrust practically every grown man in California and part of the women and children with a thirty horse-power locomotive which he might drive at will through the crowded cities, and race at express train speed over the country roads of the state, the answer would certainly have been a decided negative. Nobody on earth, I suspect, would have thought such a result possible. And yet *that is precisely what has happened*. I marvel at the success of it every time I drive in city streets. I glory in it when I see the new race of men the taxi business has created in a city like London. Contrast the clear-eyed, sober, skilful, intelligent looking London taxi-driver of to-day with the red-nosed wreck of a human being who used to be the London cabby of a quarter century ago, *and see what responsibility and power do in altering human nature.*

Also the picture which modern science has unfolded of the age-long history of the biological organism is one in which it is seen *adapting itself with marvellous success to changes in external conditions.* That we, too, at our end of this evolutionary scale, have inherited this adaptability was one of the most striking lessons of the late war in which we settled down to the endurance of what we thought intolerable conditions with amazing rapidity.

If, then, there be any notes of optimism in modern life, one of them is certainly the note played by modern science. If there be any escape from Malthusianism—the world's greatest problem—science alone can provide it. It is clearly the development of science and its application to modern life, that has made possible the support in Great Britain to-day of forty million people, when a hundred and fifty years ago Benjamin Franklin called England overpopulated with eight millions, and when Robert Fulton a little later in a prophetic mood saw England holding some time "a population of ten million souls." Perhaps this population has to-day gone too far, but the check is being applied. In both England and Sweden to-day the birth-rate is less than it is in France. *With the creative power of physical science, and the application of intelligence to the findings of biological science, even this problem of population can be faced with a good measure of hope.* An international union for its continuous study was formed this summer at Geneva. That is the objective way to begin to attack it.

Finally, can science save our civilization from the fate that has befallen its predecessors, the Sumerian, the Egyptian, the Greek, the Roman, and the others that have risen and then declined? Are the Kaiserlings, the Spenglers, and the other prophets of decay and death to be regarded as real prophets?

The answer is, of course, a secret of the Gods, but that these "prophets" are "multiplying words without knowledge" it is easy to show. For our modern civilization rests upon an altogether new sort of foundation. These older civilizations have rested upon the discovery of new fields of knowledge or of art—fields which the discoverers have indeed cultivated with such extraordinary skill that they have been able to reach a state of perfection in them that succeeding generations have often been unable to excel. Witness the sepulchral art of the Egyptians, and the perfection of such principles of architecture as they knew how to use; witness the sculpture, the painting, the æsthetic, and the purely intellectual life, of the Greeks—an accomplishment so great as to inspire an outstanding modern artist to say that there has been no advance in either sculpture or painting since the age of Pericles; witness the principles of government and of social order discovered by the Romans, or the arch in architecture, largely Roman, but reaching perhaps its perfection in the Romanesque and the Gothic of a few centuries later; witness the discovery of the principles of music in central and southern Europe in the middle ages, and the perfection that art attained within two or three centuries. And let us remember, too, that *humanity for all time is the inheritor of these achievements. This is the truth of the past which it is our opportunity and our duty to pass on to our children.*

But our modern world is distinctive, not for the discovery of new modes of expression or new fields of knowledge, though it has opened up enough of these, but for the discovery of the very idea of progress, for the discovery of the *method* by which progress comes about, and for inspiring the world with confidence in the values of that method. So long as the world can be kept thus inspired it is difficult to see how a relapse to another dark age can take place.

Even if the biological evolution of the human race should not continue—though why should what has been going on for millions of years have come to an end just now?—yet the process by which progress has been made within historic times can scarcely fail to be continuously operative. This process is the discovery of new knowledge by each generation and the transmission to the following generation of the accumulated accomplishment of the past—the discovery of new truth and the passing on of old truth.

The importance of both elements in this process has not been realized in the past and dark ages have come. But the means for the spread of knowledge, for its preservation and transmission, the facilities for universal education and inspiration, the time for leisure, and the opportunity for thought for everybody, all these have been so extended by modern science, and are capable of such further extension, that no prophecy of decline can possibly have any scientific foundation. Even arguing solely by the method of extrapolation from the past, modern science has shown that the ups and downs on the curve of history are superposed upon a curve whose general trend is upward, and it has therefore brought forth a certain amount of justification for the *faith* that it will continue to be upward. In the last analysis humanity has but one supreme problem, the problem of kindling the torch of enlightened creative effort, here and there and everywhere, and passing on for the enrichment of the lives of future generations of the truth already discovered; in a single word, the problem of *education.*

II

THE RELATION OF SCIENCE TO INDUSTRY[1]

Being, and having been all my life, an educator, I have just listened with great interest and attention to President Lowell's definition of what constitutes an education. He has phrased it a little differently, but it is almost the same definition that I heard William Allen White give in our little institution in Pasadena last year.

He was talking to our boys and he said: "You are here for an education, but you are not going to get it; you do not get any education in college, you get that afterward; all you get here is a trained mind, and if you want to know what the definition of a trained mind is, it is this: It is a mind that can listen to me for four minutes without yawning." We public speakers would like to believe that that was a correct definition.

A well-known public speaker of fifty years

[1] An address delivered at the annual dinner of the New York Chamber of Commerce in November, 1928.

once remarked ruefully after disastrous consequences had followed misplaced humor, as they
often do, "I rose by my gravity and fell by my
levity."

I use this incident as an introduction to my
speech on Science and Industry for the sake of
calling attention to the fact that what is absurd
or ridiculous to-day was perfectly good science,
or at least perfectly good philosophy, not more
than 350 years ago—that the very existence of
a "law of gravity" was discovered as late as
1650 A. D., and that "levity" and "levitation"
have through all recorded history up to Newton been just as acceptable scientific ideas as
gravity and gravitation—so recently have we
begun to understand just a little bit about the
nature of the world in which we live.

Nor do I need to go back 300 years to make
my point as to the newness of our knowledge.
It is within the memory of every man of sixty
in this audience that in the great Empire State
of New York the question could be seriously
debated, and in the most intelligent of her com-
munities, too, as to whether Archbishop Usher's
chronology computed by adding Adam's 930

years to Enoch's 365 years to Methuselah's 969 years, etc., gave the correct date of the creation. Recent election returns from Arkansas indicate that the same debate is at this very moment going on there.

But what has this to do with "Science and Industry"? Everything! For mankind's fundamental beliefs about the nature of the world and his place in it are in the last analysis the great moving forces behind all his activities. Hence the enormous *practical* importance of correct understandings. It is his beliefs about the nature of his world that determine whether man in Africa spends his time and his energies in beating tomtoms to drive away the evil spirits, or in Phœnicia in building a great "burning fiery furnace" to Moloch into which to throw his children as sacrifices to his God, or in Attica in making war on his fellow Greeks because the Delphic Oracle, or the flight of birds, or the appearance of an animal's entrails bids him to do so, or in mediævel Europe in preparing for the millennium to the neglect of all his normal activities and duties as he did to the extent of bringing on a world disaster in the year 1000,

[34]

or whether he spends his energies in burning heretics in Flanders or drowning witches in Salem, or in making perpetual motion machines in Philadelphia or magnetic belts in Los Angeles, or soothing syrups in New England.

The invention of the airplane and the radio are looked upon by every one as wonderful and pre-eminently useful achievements, and so they are—perhaps one-tenth as useful as some of the discoveries in pure science about which I wish to speak to-night, and hence worthy of at most one or two minutes of a thirty minute speech on the relations of science to industry.

As I listened in Pasadena to the presidential candidates presenting in their own easily recognizable voices from the platform in Madison Square Garden to the people of the United States their claims and the issues of the election, or at least its shibboleths, I found myself aglow with enthusiasm for the future of representative government. The few thousand citizens of Athens gathered about the Acropolis to hear the problems of the city discussed and then to cast their ballots. The 120 million citizens of the United States all had in this recent election precisely the same opportunity and in my judgment they used it judiciously. These public discussions addressed to the *ears* of the nation constitute, I think, a stupendous advance. No such step forward in public education has been taken in my judgment since the invention of printing.

But this new achievement of the race, this new capacity for education was after all only an inevitable incident in the forward sweep of pure science, which means simply knowledge, knowledge of the nature and capacities of the physical world, of the ethereal world (to which the radio belongs), of the biological world and of the intellectual world; for this knowledge, as man acquires it, necessarily carries applied science in its wake.

Look for a moment at the historic background out of which these modern marvels, as you call them, the airplane and the radio, have sprung. Neither of them would have been at all possible without 200 years of work in pure science before any bread and butter applications were dreamed of—work beginning in the sixteenth century with Copernicus and Kepler and Galileo, whose discoveries for the first time began to cause mankind to glimpse a nature, or a God, whichever term you prefer, not of caprice and whim as had been all the Gods of the ancient world, but instead a God who rules through law, a nature which can be counted upon and hence is worth knowing and worth carefully studying. This discovery which began to be made about 1600 A. D. I call the supreme discovery of all the ages, for before any application was ever dreamed of, it began to change the whole philosophical and religious outlook of the race, it began to effect a spiritual and an intellectual, not a material revolution—the material revolution came later. This new knowledge was what began at that time to banish the monastic ideal which had led thousands, perhaps millions of men, to withdraw themselves from useful lives. It was this new knowledge that began to inspire man to know his universe so as to be able to live in it more rationally.

As a result of that inspiration there followed 200 years of the pure science involved in the development of the mathematics and of the celestial mechanics necessary merely to understand the movements of the heavenly bodies—

useless knowledge to the unthinking, but all constituting an indispensable foundation for the development of the terrestrial mechanics and the industrial civilization which actually followed in the nineteenth century; for the very laws of force and motion essential to the design of all power machines of every sort were completely unknown to the ancient world, completely unknown up to Galileo's time.

Do you practical men fully realize that the airplane was only made possible by the development of the internal combustion engine, and that this in its turn was only made possible by the development of the laws governing all heat engines, the laws of thermo-dynamics, through the use for the hundred preceding years of the steam engine, and that this was only made possible by the preceding 200 years of work in celestial mechanics, that this was only made possible by the discovery by Galileo and by Newton of the laws of force and motion which had to be utilized in every one of the subsequent developments. That states the relationship of pure science to industry. The one is the child of the other. You may apply any blood test you wish and you will at once establish the relationship. *Pure science begat modern industry.*

In the case of the radio art, the commercial values of which now mount up to the billions of dollars, the parentage is still easier to trace. For if one's vision does not enable him to look back 300 years, even the shortest-sighted of men can scarcely fail to see back as much as eighteen years. For the whole structure of the radio art has been built since 1910 definitely and unquestionably upon researches carried on in the pure science laboratories for 20 years before any one dreamed that there were immediate commercial applications of these electronic discharges in high vacuum.

It is precisely the same story everywhere in all branches of human progress. I suspect it would be difficult to find one single exception. Here is the latest illustration that came to my attention less than a week ago, in fact just as I was getting aboard the train, in a letter from the Air Reduction Sales Company. It reads as follows: "We take pleasure in handing you herewith a complete set of luminescent tubes, each containing in the pure state one of the elements of the air, namely, nitrogen, oxygen, argon, hydrogen, neon, helium, krypton and xenon. It seems to us worthy of note that at the beginning of this century these gaseous elements as such had practically no commercial value or significance. To-day the estimated value of the plants and equipment that have been created either to manufacture or to use and handle these gases in industry amounts to three hundred million dollars."

The writer of this letter might have added that the chain of discovery which led up to this result started in the most "useless" of all sciences, astronomy; for helium, as its name implies, and as everybody knows, was first discovered in the sun with the aid of the spectroscope, and thirty years later it was its discovery in minute amounts in our atmosphere, also with the aid of the spectroscope, that set us looking for the other inert gases of which the letter speaks and which have recently found such enormous application in neon tubes and the like.

But why continue this recital, for no intelligent man to-day needs to be convinced that our material prosperity rests wholly upon the development of our science. It is as to the broader values, intellectual and spiritual, that even intelligent men sometimes express doubt. Let me then start with the foundations that I have already laid and try to show to what these beginnings are leading, whither we are going, not materially, but as feeling, thinking and willing beings.

Was Pasteur only a scientific enthusiast when he wrote "in our century science is the soul of the prosperity of nations and the living source of all progress. Undoubtedly the tiring discussions of politics seem to be our guide—empty appearances! What really leads us forward is a few scientific discoveries and their application."

Or was H. G. Wells, himself not a scientist at all, merely talking nonsense when he wrote quite recently (and note that he is not talking about a material thing either):

"When the intellectual history of this time comes to be written, nothing, I think, will stand out more strikingly than the empty gulf in quality between the superb and richly fruitful scientific investigations that are going on, and the general thought of other educated sections of

[41]

the community. I do not mean that scientific men are, as a whole, a class of supermen, dealing and thinking about everything in a way altogether better than the common run of humanity, but in their field they think and work with an intensity, integrity, breadth, a boldness, patience, thoroughness, fruitfulness, excepting only a few artists, which puts their work out of all comparison with any other human activity. In these particular directions the human mind has achieved a new and higher quality of attitude and gesture, a veracity, a self-detachment, and self-abrogating vigor of criticism that tends to spread out and must ultimately spread to every other human affair."

These may be extravagant statements, most of us scientists are sure they are, but I should like to attempt to picture a little of what I think was in the back of the minds of their authors when they made those statements. I shall do it by drawing an analogy between the life of mankind as a whole and the life of man as an individual. But first let me answer the question as to what we know about the duration of the life of mankind. A hundred years ago we knew practically nothing about it, as my opening remarks on Archbishop Usher's chronology showed. But since then we have made some scientific discoveries—discoveries that are not usually listed as of industrial importance at all, but which in my opinion outweigh by far, in practical value to the race, either the invention of the airplane or of the radio, and that simply because they change fundamentally our ideas about the nature of the outside world, and hence change also the nature of our acting in relation to it.

We have learned within the past half dozen years through studies in radioactivity that this world of ours has in all probability been a going concern, in something like its present geological aspects as to crustal constituents, temperatures, etc., for more than a billion years, and hence that the human race can probably count on occupying it for a very long time to come, say another billion years; and further, that mankind has been doing business on it in something like his present shape for something like 20,000 years, or possibly 50,000, but in any case a time that is negligibly small in compari-

[43]

son with the time that is behind and the time that is presumably ahead of him—in other words, we have learned that mankind, speaking of him as an individual human being, is now just an infant a few months old at the most, an infant that up to about one minute ago, for the 300 years since Galileo is but a minute in the geological time-scale, has been lying in his crib spending his waking hours playing with his fingers, wiggling his toes, shaking his rattle, in a word, in simply becoming conscious of his own sensations and his functions, waking up, as he did amazingly in Greece, to his own mental and emotional insides. Just one minute ago he began for the first time to peer out through the slats in his crib, to wonder and to begin to try to find out what kind of an external world it is that lies around him, what kind of a world it is in which he has got to live for the next billion years. The answers to that question, even though never completely given, are henceforth his one supreme concern. In this minute of experience that he has already had he has tumbled down in his crib, bumped his head against the slats, and seen stars—real ones and unreal ones, and he hasn't yet learned to distinguish with certainty between those that actually exist and those that only seem to exist because his eyeballs have been subjected to the pressure that comes from a blow, and so he is reaching out his hands part of the time trying to grasp illusions, and yet slowly, painfully learning, bit by bit, that *there is* an external world, physical and biological, that can be known, that can be counted upon, when it has once become known, to act consistently, not capriciously, that there is a *law* of gravity and that it isn't necessary to be covered with bruises all the time because he forgets that it exists, that there is a principle of conservation of energy, and that all constructive and worth while effort everywhere must henceforth take it into account and be consonant with it, that it is not worth while to spend much time hereafter with sentimentalists who wish that that law did not exist and sometimes try to legislate it out of existence, that again there are facts of heredity that it is utterly futile to inveigh against, that our whole duty is rather to bend every energy to know what they are and then to find how best to live in conformity with them, that in a single sentence there is the possibility ahead of mankind of learning, in the next billion years of its existence, to live at least a million times more wisely than we now live. This is what Pasteur meant when he said, "What really leads us forward is a few scientific discoveries and their applications." This is what Wells meant when he contrasted the result of the objective method of learning used in the pursuit of science with what he calls "the general thought of other educated sections of the community." The one guesses and acts upon its hunches or its prejudices, the other tries at least to know, and succeeds in knowing part of the time.

We need science in education, and much more of it than we now have, not primarily to train technicians for the industries which demand them, though that may be important, but much more to give everybody a little glimpse of the scientific mode of approach to life's problems, to give every one some familiarity with at least one field in which the distinction between correct and incorrect or right or wrong is not always blurred and uncertain, to let him see that it is not true that "one opinion is as good as another," to let every one understand that up to Galileo's time it was reputable science to talk about gravity and levity, but that after Galileo's time the use of levity became limited to the ridiculous, that "the town that voted the earth was flat, flat as my hat, flatter than that," had a perfect right to exist before 1400 A. D., but not after that date, that we are learning slowly, through the accumulated experience and experimenting of the centuries, especially since 1600 A. D., more about the eternal laws that do govern in the world in which we live. And for my own part I do not believe for a moment that these eternal laws are limited to the physical world either. Less than sixty years ago, to take one single illustration, there existed a relatively large political party in the United States called the Greenback Party which jumped at conclusions and which conducted campaigns to induce our government to go over to a fiat money basis. I do not suppose such a party could exist to-day unless it be in states that pass anti-evolution laws, for there are some laws that *have become established*, even in the field of finance.

[47]

This brings me to a brief discussion of the current opposition to the advance of science—an opposition participated in even by some intelligent people, on the ground that mankind cannot be trusted with too much knowledge, by others on the ground that beauty and art and high emotion are incompatible with science. Now, fear of knowledge is as old as the Garden of Eden and as recent as Dr. Faust, and there is no new answer to be made to it. The old answer is merely to point to what the increase in knowledge has done to the lot of mankind in the past, and I think that answer is sufficient, for it has certainly enfranchised the slave, and given every man, even the poorest, such opportunities as not even the prince of old enjoyed. Who would go back to the stone age because stone age man had no explosives? Of course every new capacity for beauty and joy brings with it the possibility of misuse and hence a new capacity for sorrow. But it is *our knowledge* alone that makes us men instead of lizards, and thank God, we cannot go back whether we would or no. Our supreme, our Godlike task, is to create greater beauty and fuller joy with every increased power rather than to turn our weeping eyes toward the past and fling ourselves madly, unreasoningly athwart the path of progress. Beauty in the amoeba's house disappeared when man cleaned up the miasmic swamp, but it was only because the amoeba had not the capacity to adapt itself to modern sanitation.

No, the only real question in a nation like ours is not whether science is good for us materially, intellectually, æsthetically, artistically. Of course it is, for science is simply knowledge and all knowledge helps. The only real question is how the forward march of pure science, and of applied science which necessarily follows upon its heels, can best be maintained and stimulated, for, as Pasteur said, "It is this alone that really leads us forward."

The answer to that question will depend upon the nature of one's whole social philosophy. If you think that social progress is best brought about by a paternalistic régime of some kind, by throwing upon a few elected or hereditary officials the whole responsibility for social initiative of all sorts, then you will say, "Let the government do it all; let it establish state universities and state research laboratories and state experimental projects of all kinds as it has done in most European countries and let the whole responsibility for our scientific progress lie in these institutions." But if you believe with the early makers of our nation in the widest possible *distribution* of social responsibility, in the *wide-spread* stimulation of constructive effort, in the nearest possible approach to equality of *opportunity*, not only for rising to wealth and position, but for sharing in community service, if you believe with the President-Elect that government should only step in where private enterprise fails, that it should act only as a *stimulant to private initiative* and a *check to private greed*, then your industries in the New York Chamber of Commerce, your industries which are themselves the off-spring of pure science, will join in the great nation-wide movement to keep alive the spirit of science all over this land of ours through keeping pure science going strong in the universities, its logical home, and applied science going strong in the private industrial laboratories where it thrives best. No country ever had such an opportunity as ours, no country ever had such a wide-spread stimulation of individual initiative, such a large number of citizens who had learned to treat financial power as a public trust, such resources to command, such results to anticipate. With our American ideals American industry cannot fail, I think, to realize this opportunity, and to support and keep in the finest possible condition, "the hen that lays her golden eggs." That, Mr. President, is my conception of the relation of science to industry in the United States.

III

ALLEGED SINS OF SCIENCE

The cardinal doctrine in the creed of every man of science is stated in the motto of the University of Chicago, namely, "Crescat Scientia Vita Excolatur" (let knowledge grow, let life be enriched); or equally well in the motto of the California Institute, "The Truth shall make you free." And any effort to suppress or impede the growth of science, which means to the scientist merely the growth of man's understanding of his world, and hence of his ability to live wisely in it, is to him an unpardonable sin, or at least not the work of an understanding mind.

If we are to be asked deliberately to shut our minds to the truth, or to be deterred by fear from searching for it, we might as well, so says the scientist, give up the effort at intelligent living altogether and go back to savagery. Furthermore, the whole history of man's age-long rise from superstition and ignorance up to his present estate seems to the scientist to be a practical demonstration of the essential soundness of this view.

So when a couple of years ago the Bishop of Ripon suggested to the meeting of the British Association for the Advancement of Science that it would be well for the world if science could take a ten-year holiday his words did not meet with a chorus of applause from scientists.

But the Bishop's views are not so uncommon, and we scientists have to some extent been responsible for them. The following quotation from a book written by one of the best informed and most intelligent of living Americans states the case against science thus. I quote from Mr. Raymond Fosdick's recent book entitled, "The Old Savage in the New Civilization."

"Humanity stands to-day in a position of unique peril. An unanswered question is written across the future: Is man to be the master of the civilization he has created, or is he to be its victim? Can he control the forces which he has himself let loose? Will this intricate machinery which he has built up and this vast body of knowledge which he has appropriated be the servant of the race, or will it be a Frankenstein monster that will slay its own maker? In brief, has man the capacity to keep up with his own machines?

"This is the supreme question before us. All other problems that confront us are merely its corollaries. And the necessity of a right answer is perhaps more immediate than we realize. For science is not standing still. In speaking of the scientific revolution I have not been speaking of a phenomenon that was confined to the Nineteenth Century. Rather we are just at the beginning of the revolution. We could not stop it if we would. It is advancing by leaps and bounds, gaining in impetus with each year. It is giving us more machines, faster machines, machines increasingly more intricate and complex. . . .

"Life in the future will be speeded up infinitely beyond the present. Sources of energy will be tapped and harnessed far outrivalling what we have to-day. There lies in full view before us a realm of discovery in physical science till now untrodden by mortals even in their dreams. The pioneers are already upon the road to this promised land. . . . We now know that in atoms of matter there exists a store of energy incomparably more abundant and powerful than any other of which we have thus far obtained control. If once we can liberate this force, what machines we can build! Steam and electricity will be an anachronism at which our children will laugh as we laugh at the hand loom and the spinning wheel. With a pound weight of this radioactive substance we will get as much energy as we now obtain from 150 tons of coal. Or another pound weight can be made to do the work of 150 tons of dynamite.

"One hundred and fifty tons of dynamite— enough to blow a modern city into oblivion— compressed to a pound weight which might be held in the hand! No wonder that a sober-thinking scientist like Professor Frederick Soddy of Oxford University should write: 'I trust this discovery will not be made until it is clearly understood what is involved.' 'And yet,' he goes on to say, 'it is a discovery that is sooner or later bound to come. Conceivably it might be made to-morrow.'

"One has only to turn the pages back to 1914

to find the grounds for Professor Soddy's un-
easiness. All the machines that ingenuity could
invent were directed to the single purpose of
human destruction. In a hundred laboratories,
in a thousand arsenals, factories, and bureaus,
physics and chemistry were harnessed to the
task of mass death. The gigantic success of the
enterprise is shown in the statistics: 10,000,000
known dead soldiers; 3,000,000 presumed dead
soldiers; 13,000,000 dead civilians; 20,000,000
wounded; 3,000,000 prisoners; 9,000,000 war
orphans; 5,000,000 war widows; 10,000,000
refugees.

"This was the tabulation that our mechanical civilization made possible. This is the result
of creating machinery for which we have no
method of control. This is the consequence of
giving children matches to play with. . . .

"This, then, is the problem: science will not
wait for men to catch up. It does not hold itself
responsible for the morals or capacities of its
human employers. It gives us a fire engine with
which to throw water to extinguish fire; if we
want to use the engine to throw kerosene on the
fire, that is our lookout. The engine is adapted to

both purposes. With the same hand, science gives us X-rays and machine-guns, modern surgery and high explosives, anæsthetics and poison gas. In brief, science has multiplied man's physical powers ten thousand fold and in like ratio has increased his capacity both for construction and destruction. How is that capacity to be used in the future? How can we hold in check the increasing physical power of disruptive influences? Have we spiritual assets enough to counterbalance the new forces? How can we breed a greater average intelligence? Can education run fast enough, not only to overcome the lead which science has obtained, but to keep abreast in the race? Can the old savage be trusted with the new civilization which he has created?

"These are ugly questions. They are hurled as a challenge at our generation, and upon their answers the future depends."

Now perhaps the alleged sins of science have never been stated more tellingly than in the foregoing, and I would like to ask you to allow me to bring Science herself to the witness stand and ask her what she has to say for herself.

She replies very quietly that there are both

statements and implications in the foregoing that need further consideration. First, that following her conviction that the only matter of supreme importance is to find out the facts, since we have to live with them anyway, she has kept steadily at work since Mr. Soddy raised the hobgoblin of dangerous quantities of available subatomic energies, and has brought to light good evidence that this particular hobgoblin— like most of the bugaboos that crowd in on the mind of ignorance—was a myth, that it was exceedingly fortunate that Mr. Soddy's fears did not at the time he uttered them induce a terrified humanity, like a frightened child paralyzed by its fear in the dark, to stop its efforts to get more light, for the worst disasters have always come from panic born of ignorance; that she (Science) regards it as her chief function to deter men from *over-hasty* conclusions though she does not always succeed, even with her devotees; that her influence nevertheless, is always to constrain men to replace panicky, emotional acting by reflective, informed, rational acting. The great world explosions, including the World War, have been mental, not physical. She would ask you then to withhold your judgment until all the available evidence is in.

Now the new evidence born of further scientific study is to the effect that it is highly improbable that there is any appreciable amount of available subatomic energy for man to tap; in other words, that henceforth men like the Bishop of Ripon who are living in fear lest some bad boy among the scientists may some day touch off the fuse and blow this comfortable earth of ours to star dust may go home and henceforth sleep in peace with the consciousness that the creator has put some fool-proof elements into his handiwork and that man is powerless to do it any titanic physical damage, anyway.

This may relieve the Bishop of Ripon, but it will disappoint men like Lord Birkenhead, who have been hypnotized rather than scared by the prospect of tapping enormous new sources of subatomic energy and who have been revelling in the prospect of some day lying in bed, pressing a button, and calling for two atoms' worth of massage. These men will be obliged to give up their idle Utopian dream and console themselves with the reflection that the chief joy of life after all comes from the striving and the overcoming, that there is much more satisfaction in smashing a resistant atom, as man will doubtless do, than in lying on one's back and watching it explode.

One may become blue or happy then, according to his temperament, over the fact that it is now highly improbable that we on the earth shall ever get any appreciable amounts of energy from any other source than the sun, whence we have always obtained our energy, directly or indirectly, in the past, but at any rate that is the indication to which we must adjust ourselves, and it serves at least to remove from the account of Science one sin with which she had been charged.

But that is but the first of the sins charged against her. What about the horrible indictment as to the twenty-six million people actually killed in the world war? The answer is twofold. First, the implication was that Science had a good deal of responsibility for that war,—an erroneous implication I think, since war has been the chief business of all the glorious civilizations of the past when there was no science, and with every advance in science I think it becomes less and less so. Indeed, primitive man's chief tools were probably arrowheads and tomahawk and his chief industry making and using them. When the age of bronze replaced the age of stone a multitude of new peaceful arts were born. Coppersmiths, silversmiths, goldsmiths appeared who developed a wonderful decorative art for use on urns, on vases, on table ware, on personal ornaments, on sarcophagi, on friezes, on monuments,—witness the amazing perfection of these arts revealed in Tutankamen's tomb,—*and these arts reduced the relative importance of the successor of the arrowhead and tomahawk maker,* for these peaceful arts turned men's minds and energies and interests away from war, toward peace.

And this has been the consequence I think of practically every advance in science and its applications since that time. Let him whose eyes have recently been focussed on the increased effectiveness of tools of destruction and whose fears have been aroused lest the savage in man may use these tools to destroy the race lift up his head and look all around him on all sides. I think that such a survey will show conclusively that *every scientific advance finds ten times as many new, peaceful, constructive uses as it finds destructive ones.* Explosives and fertilizers are basically the same, and even explosives as such meet a dozen peaceful needs to one warlike one. The duPont Company is known as a powder concern, but that is a well-nigh negligible part of its business. Public thinking is misled merely by the fact that a horror makes better news than a wheat crop. One man blown painlessly to atoms gets more news space than a thousand men in the agonies of starvation or dying by inches from disease. Steel does indeed make bayonets, but it also makes plowshares, and railroads, and automobiles, and sewing machines, and threshers, and a thousand other things, whose uses constitute *the strongest existing diverter of human energies from the destructive to the peaceful arts.*

In my judgment war is now in process of being abolished chiefly by this relentless advance of science, its most powerful enemy. It has existed in spite of religion, and in spite of philosophy, and in spite of social ethics, and in spite of humanitarianism and the golden rule, since the days of the cave-man because, in accordance with the evolutionary philosophy of modern science, and simply because, *it has had survival value.* It will disappear like the dinosaur when, and only when, the conditions which have given it its survival value have disappeared, and those conditions are disappearing now primarily because of the changes in world conditions being brought about by the growth of modern science.

I am with Mr. Fosdick in every effort to arouse more fully the social sense, the conscience, and the morals of mankind, in every effort to develop a new machinery like a world court and a League of Nations to assist in bringing about better international and social relations. If I differ with him at all, and I am not sure that I do, it is only in my estimate of the relative effectiveness of the different available agencies. He seems to fear too active experimenting in physics and its applications, but not in sociology, for when he is dealing with the latter field he says, "We need not fear that we shall progress too fast. The overwhelming danger is that we shall not be able to progress fast enough."

My own reason for exactly reversing this emphasis is very precisely stated by him in the following words: "But social science to-day is still lacking in the fundamental groundwork of knowledge. It is still too largely based upon inspiration rather than upon facts." My own position stated in one sentence is that all progress comes from *knowledge*, and I am enthusiastically for everything that increases knowledge, whatever be the field, sociology or physics, and for acting upon that knowledge when found. But it is as unsound to talk about the danger of too much knowledge in physics as in sociology.

For look further at what is actually happening, at what kind of effort is to-day yielding the largest *social* returns. Without aiming directly at doing so, modern science and its applications have within the past fifty years actually produced the most profound and beneficial social changes that the world has ever seen. They have raised the average working man's wage in terms of actual buying power about fifty per cent in forty years, and that along with a twenty per cent decrease in his working hours. That is not everything, but *it is the necessary first step,* the indispensable foundation upon which all other kinds of building must rest. Also, according to recent "Carnegie" studies modern science and its applications have actually resulted in increasing the amount of reading done by the average man in Middletown more than three hundred per cent, and this applies, too, to magazines like the *Atlantic Monthly* as well as to other grades of reading matter. Also, they have given the average man through the radio and the "movie" *the opportunity* for education and entertainment (partly abused, no doubt, but partly utilized; and in any case opportunity is a sine qua non to progress)—opportunity such as the common man never had before.

One continually hears complaints that our machine age, with its mass production, has ruined the life of the common man, that it has deadened and routinized labor and taken away the joy of craftsmanship. These protests are natural, because the man who is taken through

[65]

a modern factory and does not look beneath the surface of things will easily gain such an impression. A very superficial glance at the Ford factory, for example, would seem to justify the worst charges that are made against our machine age, but to the man who is capable of seeing beyond his nose it is a very different picture that unfolds itself. This man sees not merely the 8,000 cars turned out each day by routine labor in the summer of 1929 by the Ford plant, but he looks beyond to see *what these cars are doing to the life of the common man.* He sees, in the first place, these eight thousand cars driven by roughly as many persons, and he realizes that driving a car in crowded streets, which is in itself a highly skilled occupation, develops in large measure the qualities of sobriety, alertness, and intelligence. He contrasts the bleary-eyed, ruby-nosed old soak who thirty years ago sat on the driver's seat of the average cab in London or New York with the highly skilled chauffeur of to-day, alert, self-respecting, sober, intelligent, and well dressed. *The change is striking and the improvement enormous.*

Secondly, he sees that every one of those 8,000 cars turned out in one day has to be taken care of and repaired by an intelligent garage mechanic and "trouble man," a man who must understand the intricate mechanism of an automobile from top to bottom and from one end to the other, who must be able to find the difficulty no matter where it appears, and, more than that, *who must know how to right it*. No such requirement of expert knowledge, no such variety of stimulating activity, was ever the lot of any middle-age artisan, and no such demand for intelligence was ever made before upon so large a fraction of the population as is now made by our mechanical age.

Thirdly, he sees that all these 8,000 cars, and literally millions more, must be serviced by thousands of wide-awake, courteous, attractive service-station men—men who have taught the world as it has never been taught before that the maximum of success is definitely related to the maximum of cheerfully rendered service to one's fellows. He sees that back of these service-station men are the refineries, with their expert staffs of chemists and physicists, and that back of these are the geologists and the seismologists and the radio engineers of the producing company, and so on without end.

As I read history, the machine age has actually freed, educated, and inspired mankind, not enslaved it! Routine labor plays a part in all our lives, and an attractive part too, if it is not overdone, and if there is leisure for something else. Witness the endless repetition involved in knitting, in engraving, in practically all so-called skilled occupations. Even the few routine men who feed the machines in Mr. Ford's factory are less routinized and have shorter hours than the dumb agricultural drudge who hoed potatoes for twelve hours a day through all the history of the world before the machine age appeared.

But the far-seeing man will see even deeper than that. It is science and its applications that, through the Ford car and its like, have given to the average man and his family the opportunity for the broadening influences of travel, an opportunity that he is utilizing amazingly, too. It is science and its applications that, through the wonderful development of the art of communications, and through incredible stimulation and acceleration of trade and commerce, have knit the whole world together into a unity that makes war an anachronism. Much more important than treaties, I think, in abolishing war as an instrument of national policy is the growing recognition of the fact, taught in no uncertain language from 1914 to 1918, that in our modern scientific civilization war is no longer well adapted to the attainment of national ends. Let no man henceforth ever make the error of assuming that modern science made the last war. Rather was that war, let us hope, the last titanic struggle of militarism to escape the extinction foredoomed for it in a world motivated by modern science. The world war was surely not a sin of science.

But, yet, can science escape the responsibility for those twenty-six million lives lost during that war? That is a matter of opinion. Granted only that these people could all have got to the fighting line, which, mind you, was of wholly unprecedented length, I am not sure that, with only ancient man's weapons, the sword, the shield, and the spear, given the

[69]

world's war issues to fight about, the thirteen million who died in battle might not have been even more; and without modern medical science the thirteen million civilian deaths would almost certainly have been augmented.

But that is after all not particularly important. The significant fact is one brought out by Mr. Fosdick himself when he says, "Stop the machines and half the people in the world would perish in a month." That is not an over-estimate. Modern science undoubtedly made it possible for more than twice as many people to live comfortably in Europe before the war as could otherwise have done so. Robert Fulton predicted in about the year 1800 A. D. that the time would "come when England, then thought densely populated, would hold 10,000,000 souls." To-day she has five times that number. It was but a small fraction of these people, people who owed their very existence to science, who had been created by science, that lost their lives in the war. Had preceding generations abolished or slowed up science, more than this number would have died more miserably, *i. e.*, with greater suffering, for disease with science is bad enough while without science it is hell.

Now the balance of this whole account scarcely shows a sin to be credited to science. Looked at in the large, I do not think there can be the slightest question that the only hope that this world has of maintaining in the future a suitable balance between population and the food supply is found in science. That, in the last analysis, is mankind's greatest problem. Its solution alone, and there are the best of reasons for believing that in the long run it can be solved, is sufficient to warrant the fullest stimulation of both the biological and physical science that can in any way be brought about.

So far in my search for the sins of science I have failed to find her guilty of the charges brought against her, but there is one to me very regrettable tendency in modern life for which science is probably to some extent at least responsible. I refer to the craze for the new regardless of the true, to the demand for change for the sake of change, without reference to the consequences, to the present-day wide-spread worship of the bizarre, to the cheap extravagance and sensationalism that surround us on every side, as evidenced by our newspapers, our magazines, our novels, our drama, our art in many of its forms, our advertising, even our education.

These are, I suppose, inevitable, though I hope transient, accompaniments of the stupendous *rate of change* that modern science and its applications have forced on modern life. *The spirit of change has been caught where its basis has been wanting.* In this particular our generation stands unique in all history, and it is difficult to see how the future can have any other period of so rapid change in store.

In the way he conducted his daily life, and also in his prejudices and superstitions, my grandfather is undoubtedly more remote from me than was he from the earliest man mentioned in recorded history. About 1830 he took his Lares and Penates, his flocks and his herds, and moved west from Massachusetts to Illinois by foot and by horse in precisely the way in which four thousand years earlier Abraham had moved westward from Ur of the Chaldees. When I myself went to New York to study in 1895, no ship ever sailed on Friday, for the actions of men were still governed, as they had been for a thousand years and more, by that particular superstition. To-day all that is changed! In the last analysis this change is primarily due to the introduction of the power machine as a substitute for animal muscle, for this includes everything that has come to this generation through the steam engine, the dynamo, the automobile, the airplane, the telephone, and the radio. Add to this the change in *mode of thought* due to the new host of discoveries, primarily in physics and biology, and it is no wonder that our age has become infected, or better drunk, with the spirit of change. *In many fields no past time has known and no future time can know so* sudden and so complete a transformation, for the whole gamut of possibilities has been run through by our single generation. In woman's dress, for example, the limits are obviously zero and infinity, and whatever there is in between, that has not been tried since 1900, isn't likely to be tried very soon, nor would it represent a very large change if it were, so that whatever zest and joy there be in something brand new and radically different in this domain has been tasted to the full by this generation, and will never be tasted in such completeness again.

In physics and its applications these changes have been made by men who were fully conversant with the past, men who knew the difference between perpetual-motion cranks and real discoverers, men who knew that *the fully verified laws of the past must remain the laws of the future for the whole range of phenomena for which their correctness has been tested;* in a word, men who knew that Einstein would have to contain the whole of Newton, *i. e.*, be merely a refinement of, and supplement to, Newton, or else that his work would be wrong. But unfortunately many of the other fields in which the spirit of change is rife have no such criteria for past or present truth as physics possesses and no such group of well-trained, capable, and historically informed minds working in them, so that in these fields we cannot be certain whether the changes represent progress or retrogression. In such cases, however, the counsel of the wisest heads of the past is the only possible guide for the present.

But be that as it may, I suppose that the present spirit of revolt, of change for the sake of change, the present effort for the new at all

costs, the bizarre, the extravagant, the sensational, is in part an inevitable reflex of the rapid
changes taking place in our times because of the
rapid growth of science. When I go into an ex-
hibition of the so-called secessionists in art in
Germany I feel certain that I am in a madhouse,
or when I read the literature poured forth by
what Mr. Stuart Sherman called the emetic
school of modern American writers, I dislike to
admit that these modern excrescences of our civilization are a part of the sins of science, but I
suspect the spirit of change which we have
started has been partially responsible for them.

But I am not greatly disturbed even by these.
The world will become sick of the emetic school
in time. The actual method by which science
makes its changes is becoming better understood.
The demand for the saner popular books upon
it is continually increasing. The remedy is in
part at least in understanding it better.

As soon as the public learns, as it is slowly
learning, that science, universally recognized as
the basis of our civilization, knows no such thing
as change for the sake of change, as soon as the
public learns that the method of science is not

to discard the past, but always to build upon it, to incorporate the great bulk of it into the framework of the present, as soon as it learns that in science truth once discovered always remains truth, in a word that evolution, growth, not revolution, is its method, it will I hope begin to banish its present craze for the sensational, for the new regardless of the true, and thereby atone for one of the sins into which the very rapid growth of science may have tempted it.

But there is another side even to this admitted sin which will appeal to those of us who want to speed up social change, to those who feel that many of our laws and customs have actually become outgrown, that they were developed for, and were adapted to, the old civilization, not to the new. In many, many instances this view is undoubtedly correct. But here the sin just now admitted becomes a virtue. That the spirit of change is in the air obviously helps rather than hinders in the case of these needed social readjustments. The whole question however is, "do we know enough yet to make any particular change?" The answer is sometimes yes and sometimes no.

But in the latter case the new knowledge that is still needed is just as likely to come from further physical experimenting as from further social studies. The whole history of science shows that it is impossible to predict beforehand where a new bit of knowledge is going to fit in. The amazing thing about that history is the extraordinary rapidity with which each new advance in one domain actually finds its application in another. Physical knowledge *is* social knowledge! Let us not then hold back anywhere in the search for knowledge. Crescat Scientia, Vita Excolatur.

There is one other sin that is charged against science concerning which I wish to say a word, namely the alleged sin of exalting the material at the expense of the spiritual.

If this means providing food and clothing and wholesome living conditions for millions upon millions of people who would otherwise die of starvation or otherwise drag out so miserable lives that their only recourse would be to dream of another life free from the miseries of this, then science must plead guilty.

The rise of science has undoubtedly filled mankind with a new vision of, a new hope for,

[77]

and a new effort toward a better human existence than the world has known in the past. If this is exalting the material over the spiritual then she must again plead guilty.

The rise of science has undoubtedly shifted somewhat the relative emphasis of our thinking from individual-soul-salvation to race salvation. If this is exalting the material, then she is again "guilty."

But as I myself use words the foregoing facts do not mean the subordination of the spiritual to the material. I myself think that the aforesaid changes represent an increase rather than a decrease in what I call *"spiritual values,"* *i. e.,* an increase in the essential spirit of the great teacher which was epitomized in the Golden Rule. The essence of Christianity is to be found, I think, in the altruistic teaching and living which Jesus felt it to be his chief mission to spread on earth. I have no reason to think that this spirit is on the wane. Even the membership in the christian churches, which are the chief stimulants of it, is increasing, and a civilization built upon modern science unquestionably demands its further increase. For as society becomes more and more complex civilization cannot endure at all save as the individual learns in ever-increasing measure to subordinate his own appetites and impulses to the common good, to the group life wherever the two come into conflict. In other words, the development of the sense of social responsibility which, broadly speaking, is merely the spirit of the Golden Rule, or slightly differently stated, the stimulation of the "consciences, the ideals and the aspirations of mankind," must be done in ever-increasing measure in a civilization which is growing more and more complex and interrelated under the influence of modern science.

So much for the practical side of the question. But there is also a philosophic side. Science is sometimes charged with inducing a materialistic philosophy. But if there is anything which the growth of modern physics should have taught it is that such dogmatic assertiveness about the whole of what there is or is not in the universe as was represented by nineteenth-century materialism is unscientific and unsound. The physicist has had the bottom knocked out of his generalizations so completely that he has learned with Job the folly of "multiplying words without knowledge" as did all those who once asserted that the universe was to be interpreted in terms of hard, round, soulless atoms and their motions. Says the Oxford biologist, John Scott Haldane, "Materialism, once a scientific theory, is now the fatalistic creed of thousands, but materialism is nothing better than a superstition on the same level as a belief in witches and devils. The mechanistic theory 'is bankrupt.'"

The best possible cure for materialism is the following chapter from the recent history of physics. A hundred years ago, physics consisted of six distinct, sharply separated departments, Mechanics, Molecular Physics, Heat, Sound, Light, Electricity. The first partition between these compartments to be broken completely down was that between heat and molecular physics, when about 1850 heat was found to be not a substance, as had been supposed, but simply molecular motion. The next discovery was that radiant heat and light were not different categories of phenomena but essentially the same phenomenon, that they were both ether waves identical save for wave length. The next great discovery, made by Maxwell and Hertz, was that electric-wave phenomena are indistinguishable from light and heat save for wave length. All these phenomena of radiant heat, light and electric waves then became fused under the general heading "ether-physics," still sharply separated from matter-physics and also from current electricity.

The next partition to go was that between current electricity and matter-physics, when electric currents were found to be the motions of electrons. But one partition then remained, that between ether-physics and matter-physics. Quite recently this too is gone, and matter and ether waves are fused together in Einstein's Equation and ether and matter become indistinguishable terms. Electrons are now both particles and waves, and light waves are also corpuscles. What does it all mean? Simply that *there is an interrelatedness, a unity, a oneness about the whole of nature,* and yet still an amazing mystery. Is it at all likely in the light of that history that we can long maintain airtight compartments separating ether (or matter, whichever you will) from life and mind?

Now another finding of modern physics! With astonishing rapidity within the past twenty years man has extended his vision. He has looked inside the atom, a body one millionth the diameter of a pin head, and found an infinitely small nucleus one ten-thousandth the diameter of the atom and arranged about it as many as 92 electrons (in uranium) each playing its appropriate rôle in a symmetrical, co-ordinated atomic system. He has then looked inside that nucleus and counted in uranium exactly 238 positives and 146 negatives, and he has found that *the atom changes to something else if any one of these positives or negatives drops out.* He has watched the interplay of radiation upon these electrons, both within the nucleus and out of it, and found everywhere amazing orderliness and system. He has learned the rules of nature's game in producing the extraordinarily complicated spectrum of a substance like iron, for example, and it is, in Sommerfeld's phrase, unbelievable zauberei (magic) that these complicated rules never fail to predict exactly the observed results. Again, man has turned his microscope upon the living cell and found it even more complex than the atom, with many parts each performing its function necessary to the life of the whole, and again he has turned his great telescopes upon the spiral nebulæ a million light years away and there also found system and order.

After all that is there any one who still talks about the materialism of science? Rather does the scientist join with the psalmist of thousands of years ago in reverently proclaiming "the Heavens declare the glory of God and the Firmament sheweth his handiwork." The God of Science is the spirit of rational order and of orderly development, *the integrating factor in the world of atoms and of ether and of ideas and of duties and of intelligence.* Materialism is surely not a sin of modern science.

I have thus found science "not guilty" of most of the specific counts raised against her. But after this defense I am ready to go back to the quotation from Mr. Fosdick and join him in raising precisely the question he there asks. For in the last analysis that question is merely whether for any reason whatever, scientific or non-scientific, mankind, or more specifically this particular generation of Americans, has the moral qualities that make it safe to trust it with the immensely increased knowledge and the correspondingly increased power which has come into its possession. I join him in throwing out that question as a challenge to our generation, for there can be no doubt that our generation has been getting hold of the sources of knowledge and of power at a rate such as no generation of the past has ever known, and so far at least as mechanical power is concerned such as no generation of the future is likely to know.

I am not in general disturbed by expanding knowledge or increasing power, but I begin to be disturbed when this comes coincidently with a decrease in the sense of moral values. If these two occur together, whether they bear any relationship or not, there is real cause for alarm.

Now there are certain disturbing indications in America just now of such a coincidence. I will mention but two of them: the one is the obvious effort at *the deflation of idealism,* the ridiculing of the existence of such a thing as *a sense of duty* or of social responsibility, not,

[84]

thank God, by scientists; but rather by a group of American writers which is apparently trying to create something brand new in morals; and the second is the apparently increasing lawlessness just now characteristic of American life. When we have now, and have had for twenty years, *i. e.*, for a time long antedating the advent of prohibition, *sixteen times the number of homicides per thousand of population that are found in England* there is some reason for alarm. Where individuals in sufficiently large numbers are willing to destroy the basis of confidence in themselves by refusing to be governed by the rules which they themselves, with the aid of their recognized and duly established and agreed upon machinery, have set up, then obviously the foundations of civilization are being completely undermined. If that spirit coexists with the destructive possibilities brought forth by modern science the danger is very great. The remedy, however, is obviously not to try to hold back the wheels of scientific progress, but rather to use every available agency, religious, social, educational, as individuals, as groups, and as a nation, to stay the spread of the

[85]

spirit of selfishness, lawlessness and disintegration. That, I take it, is essentially the challenge of Mr. Fosdick's book, and in that challenge I am quite ready to join with him.

IV

AVAILABLE ENERGY[1]

Astronomy is often called the most useless of the sciences, and so it is from the standpoint of the man whose time-horizon extends ten years forward and ten years backward; and that man, too, probably represents ninety per cent of all mankind. But for the smaller fraction of men who have been able to rise above the mole's outlook, who have studied enough of the past and understood enough of the present to have acquired a three hundred year time-horizon, for all such the foregoing statement is grotesquely incorrect.

If utility consists in nothing more than feeding and clothing the generation now living, then there are indeed *useless* sciences, and astronomy is perhaps one of them. But again, if utility consists in nothing more than feeding and clothing ten successive generations, then, even by that narrow standard, the verdict of history has

[1]Address made in September, 1928, in response to the presentation of the gold medal of the British Society of Chemical Industry.

definitely been that astronomy is one of the most useful of all the sciences, more useful probably than physics, chemistry, geology or engineering, and that for the simple reason that without it there would presumably never have been any modern physics, chemistry, geology or engineering. Eliminate it and you probably eliminate with it the development of the whole of Galilean and Newtonian mechanics: you certainly eliminate the discovery of the law of gravity, and of all the principles of *celestial mechanics*, and you probably eliminate even the discovery of the laws of force and motion. All these discoveries which came out of the labor and travail of two long centuries, the seventeenth and the eighteenth, which had to create even a new mathematics in order to be able to handle the new group of physical ideas, constituted an indispensable foundation for the crowning achievement of the nineteenth century, namely, the *application of these same laws to the development of terrestrial mechanics,* an achievement out of which has grown most, if not all, of the *distinctive* features of *modern* civilization.

Utility can only be properly defined as all that contributes to the finer, fuller, richer, wiser, more satisfying living of the race as a whole, and there is scarcely a bit of knowledge of the external world, or of man himself, that does not definitely help toward that end. Some of the most useful discoveries have exerted their chief influence, not through showing how the yield of beans or of cabbages per acre could be doubled, but rather through preventing mankind from wasting its precious energies on useless undertakings, such for example as building a tower of Babel. All knowledge that helps toward an understanding of the nature of the universe of which we are a part is useful, for we need very much more of it than we now have, or shall have for centuries to come, to enable us to direct our energies toward wise, effective living instead of wasting them on beating tom-toms, inventing perpetual-motion machines, or chasing either physical or social rainbows. "A penny saved is a penny earned," and this is quite as true of human energies as of household economics.

The disasters that can befall mankind merely

because of erroneous conceptions of the nature of the world in which we live are well illustrated by the historic record of the miseries that came upon the earth in the year 1000 A. D. because of the wide-spread belief that the world was coming to an end at that time. The recent exact measurement of the amount of lead in Black Hills uraninite, and of the exact atomic weight of that lead, is not usually regarded as a great engineering undertaking, nor as an accomplishment fraught with important useful consequences; but I venture the estimate that the knowledge that has come from that and similar experiments to the effect that this world has already had a lifetime of at least a billion years, and that man has in all probability another billion years ahead of him, in which there is the possibility of his learning to live at least "a million times more wisely" than he now lives, is likely to have in the long run a much larger influence upon human conduct than the invention either of the airplane or of the radio, important and pre-eminently useful though these be. Similarly, the discovery of the second law of thermodynamics has been perhaps more useful in preventing useless effort, in spite of the legions of perpetual motion cranks who still infest all physical laboratories, than in improving the efficiency of the heat engine.

The foregoing considerations constitute my argument for the appropriateness of presenting before the Society of Chemical Industry some discoveries that have not heretofore been labelled "useful" but which, when put together and correctly interpreted, as I hope they are herewith, may constitute an important and perpetually burning beacon to point out to mankind the way of progress, even of *industrial* progress.

The four recent developments in the field of pure science that I am herewith endeavoring to fuse together into a result of industrial importance are: (1) the discovery of the relation between mass and energy; (2) the development of methods of making very exact atomic-weight determinations; (3) the discovery of the cosmic rays; (4) the development of relativity-quantum mechanics. It may appear at first sight to the average supporter of Al Smith that I have taken a large order, but pray withhold your judgment.

[91]

The first of these four very recent developments was stated in its general equational form, namely, $mc^2 = E$, in which m represents mass in grams, c the velocity of light, and E energy in ergs, by Einstein as one of the most important consequences of the special theory of relativity (1905); but it had been experimentally established for special cases before Einstein's day, namely, in 1901 and 1902 by Kaufmann's measurements on the variation of the mass of the electron with its kinetic energy, and it had also been made by Lorentz a theoretical consequence of the electro-magnetic theory of the origin of mass. This equation may therefore, I think, be taken as a safe practical guide even by those who hope that the special theory of relativity may ultimately be found to involve some second-order error. The Michelson-Morley experiment has surely been sufficiently checked to establish the fact that it involves no first-order uncertainty.

The second of the foregoing developments is due primarily to Aston, of Cambridge University, England, who last summer, 1927, established experimentally a definite relation represented by a smooth curve, between the atomic weights of the elements and the mass of the positive electron as it appears in the constitution of the nucleus of each particular atom. This relation is a purely empirical one, the bearing of which upon the argument herewith presented has never, so far as I know, been pointed out before, at least in a quantitative way, save in very recent papers by Dr. Cameron and myself, the most complete and important of which is to appear in the October number of *The Physical Review* (1928).

The point upon which we lay emphasis is that if this smooth experimental curve may be taken as a safe guide, then by combining it with Einstein's equation we can at once draw very important conclusions about the possible sources of available energy.

The first conclusion that we draw is that *the process of radioactive disintegration with the ejection of an alpha particle is a process that can take place only in the case of a very few of the very heavy and very rare elements.* For radioactivity is a heat-evolving, *i. e.*, an exothermic process, otherwise it could of course not take place of its own accord, and Einstein's equation tells us that no energy-evolving or exothermic process can take place unless the total mass of all the constituents after the change is less than the total mass before the change—that an equivalent mass must always disappear if other forms of energy are to appear. But the relation of these masses before and after any hypothetical change is just what is given by Aston's curve, which shows that no element of atomic weight under say about 100 can disintegrate with the ejection of an alpha particle and the evolution of energy. And yet more than ninety-nine per cent of all matter consists of these atoms of atomic weight less than 100. *Therefore radioactivity with the ejection of alpha particles, even of a very feeble energy,* is not a general property of matter, as many of us have in the past thought it might be. Under the stimulus of the discovery of the enormous quantities of energy evolved in the disintegration of uranium and thorium we have often imagined, and sometimes incautiously stated, that there might be similar amounts of available energy locked up in the common elements, releasable, perchance, by getting them to disintegrate, as uranium and thorium spontaneously are doing. And engineers, physicists and laymen alike have talked glibly about "utilizing this source of energy when the coal is gone." So-called humanists, on the other hand, advocates of a return to the "glories" of a pre-scientific age, have pictured the diabolical scientist tinkering heedlessly, like the bad small boy, with these enormous stores of sub-atomic energy, and some sad day touching off the fuse and blowing our comfortable little globe to smithereens.

But Nature, or God, whichever term you prefer, was not unconscious of the wisdom of introducing a few fool-proof features into the machine. If Einstein's equation and Aston's curve are even roughly correct, as I am sure they are, for Dr. Cameron and I have computed with their aid the maximum energy evolved in radioactive change and found it to check well with observation, then this supposition of an energy evolution through the disintegration of the common elements is from the one point of view a childish Utopian dream, and from the other a foolish bugaboo. *For the great majority of*

[95]

the elements, such as constitute the bulk of our world, are in their state of maximum stability already. They have no energy to give up in the disintegrating process. They can only be broken apart by working upon them, or by supplying energy to them. Man can probably learn to disintegrate them, but he will always do it "by the sweat of his brow."

But having thus disposed of the process of atomic disintegration, and found it completely wanting as a source of available energy, since the radioactive elements are necessarily negligible in quantity, let us next see what there is to be learned about the process of *atom-building* as a source of energy. Here Einstein's equation, Aston's curve, and the third of the foregoing developments in pure science, namely, the recent experimental work on cosmic rays, have just thrown a flood of light on the processes going on in this universe in which we live. For, first, Dr. Cameron and I have recently found three definite cosmic ray bands, or frequencies, of penetrating powers, or ray-energies, respectively, about twelve, fifty and one hundred times the maximum possible energies that are, or can be, obtained from any radioactive, that is, any disintegrating, process. The highest frequency band has so enormous a penetrating power that it passes through more than 200 feet of water or eighteen feet of lead before becoming completely absorbed, while two or three inches of lead absorbs the hardest gamma rays. This discovery of a banded structure in cosmic rays shows that these rays are not produced, as are X-rays, by the impact upon the atoms of matter of electrons that have acquired large velocities by falling through powerful electrical fields, as we earlier suggested—the fields needed to produce frequencies as high as those of the highest observed cosmic rays are equivalent to 216,-000,000 volts—but that *they are rather produced by definite and continually recurring atomic transformations involving very much greater energy-changes than any occurring in radioactive processes.*

Taking Einstein's equation and Aston's curve as a guide there are no possible atomic transformations capable of yielding rays of the enormous penetrating power observed by us, except those corresponding to the building up or crea-

tion of the abundant elements like helium, oxygen, silicon, and iron out of hydrogen, or possibly in the case of the last two elements out of helium. The entire annihilation of hydrogen by the falling completely together of its positive and negative electrons has been suggested as an additional possibility, but it can be eliminated in this case for two excellent reasons. The first is that there is no place for such a radiation to occupy in the observed cosmic-ray curve; tested by any formula that has ever been suggested by any one to relate energy to absorption coefficient (the so-called Klein-Nishina or the Dirac), this annihilation act should produce one single big radiation band *just where there is none;* and the second is that this radiation, if it were present, would necessarily be homogeneous and could not by any possibility exhibit the type of multiple bands definitely shown by the observed cosmic rays. So that here alone, by a process of exclusion, we have arrived at pretty definite evidence that *the observed cosmic rays are the signals broadcast throughout the heavens of the births of the common elements out of positive and negative electrons.*

But right here is where the fourth of the aforementioned recent developments in the purest of pure science dovetails into the practical picture. Dirac is a young Englishman, Klein a young German, and Nishina a young Japanese, all deeply versed in what the engineer of to-day—and even the experimental physicist too—is rather proud to say "is altogether beyond him," meaning thereby that he considers that there are more important things for him to put his energies upon. Such highbrow subjects as relativity-quantum mechanics, and the new wave-mechanics, to which these men have made outstanding contributions—where do they touch life anyway? Very quickly has come the answer. Dr. Cameron and I had measured approximately the penetrating powers, or absorption coefficients, of our three prominent cosmic-ray bands. Without being guided by any theory at all we first found them at $\mu = 0.35$, $\mu = 0.08$, $\mu = 0.04$ where μ means absorption coefficient per meter of water. The foregoing gentlemen's formulæ, giving the quantitative relation between absorption coefficient and frequency, or energy, these formulæ all being extensions of

[99]

and corrections to one originally worked out with consummate skill by Arthur H. Compton, of Chicago, last year's Nobel Prize winner in physics, enable us to compute from Einstein's equation and Aston's curve what should be the absorption coefficients, or the penetrating powers, of ether-waves produced by the act of creation of the common elements out of the primordial positive and negative electrons.

Before presenting these computations, however, let me build a little more background. It is an interesting and a very important fact from the practical view-point, too, that more than ninety-five per cent of this universe, so far as we can now see, is made up of a very few elements.

First. The spectroscopy of the heavens shows the enormous prevalence everywhere of hydrogen, but hydrogen is merely the primordial positive and negative electrons tied together, or in process of being so tied.

Second. The spectroscopy of the heavens also shows that helium is an exceptionally abundant, and a very widely distributed, element, even though, because of its lightness and inability to

[100]

combine with anything, even with itself, the earth has not retained much of it. Significant is it, however, that the alpha particle given off by all the heavy radioactive elements is nothing but helium, so that it must have a certain prevalence even on earth in the structure of the heavier elements.

Third. Dr. I. S. Bowen, at the California Institute, solved, with consummate skill, the half-century-old riddle of "nebulium," and has shown that the substance, abundant throughout the heavens, giving rise to these mysterious spectral lines, is mainly oxygen and nitrogen. But oxygen alone constitutes 55 per cent of the earth's crust, and about the same proportion of meteorites. Oxygen and nitrogen, then, which for our present purpose will be treated as one element, since they have nearly the same atomic weight and will be henceforth listed under the name of the stronger brother, constitute the third extraordinarily abundant element, and it is to be noted that there are no abundant elements at all between helium and oxygen. Carbon has a certain minor prevalence, but because of its nearness in atomic weight to nitrogen and oxygen it may here be treated as merely a feeble satellite to oxygen.

Fourth. Ninety-five per cent of the weight of all meteorites consists of oxygen (54 per cent), magnesium (13 per cent), silicon (15 per cent) and iron (13 per cent), while 76 per cent of the earth's crust is composed of the three elements, oxygen (55 per cent), silicon (16 per cent) and aluminum (5 per cent), no other element rising over 2 per cent. Iron constitutes 1.5 per cent of the crust, but it is supposed to be very largely represented in the interior. Because of the closeness in their atomic weights magnesium, aluminum and silicon (24, 27, 28) may, for our present purpose, be regarded as a single element and given the name of the strongest brother, silicon. There are then no abundant elements whatever between oxygen and silicon, nor between silicon and iron (atomic weight 56), *and from an engineering standpoint the universe may be said to be made up of the primordial positive and negative electrons, and of four elements built out of them, namely, helium, oxygen, silicon and iron.*

Let me now digress from my subject, "Available Energy," just long enough to point out the practical significance of the foregoing facts. Mankind, if he is here a billion years hence, will be satisfying his main needs, as he satisfies them now, with the four elements, hydrogen, oxygen, silicon and iron, *i. e.*, with water, air, earth and Fe, where the last symbol stands for iron rather than for fire, which was the fourth constituent of the world of the ancients. These fundamental facts may some time help to stabilize the stock market. Aluminum might some day compete with iron in usefulness if lightness were a desideratum, but for the great bulk of structural purposes it is not. The progress of science and invention is not likely to put out of business for a billion years to come the concerns engaged in the iron and steel industry.

But my subject to-day is not available materials but rather available energy. Einstein's equation and Aston's curve, then, enabled Dr. Cameron and myself to compute the energies sent out in the ether signals arising from the creation in single isolated acts of helium, oxygen, silicon and iron, and then Dirac's formula enabled us to compute the absorption coefficients of the corresponding cosmic rays. The theoretical values of the absorption coefficients corresponding to the first three of these creative acts came out $\mu = 0.30$, $\mu = 0.08$, $\mu = 0.04$, as compared with the previously obtained and already reported observed-values $\mu = 0.35$, $\mu = 0.08$, $\mu = 0.04$. The agreement is much better than our observational uncertainty,[1] and leaves little doubt in our own minds that *the observed cosmic rays are in fact the birth-cries of the infant atoms of helium, oxygen and silicon.* We have some little indications that we can also hear the shriller birth squeaks of infant iron, though this is a little more uncertain.

But the question that is already being asked on all hands is "Where are these atom-building processes going on?" To this question, too, we think we have the answer. It is "not at all in the stars," for high temperatures and densities

[1] Indeed this quantitative fit is significant only as to order of magnitude. The fact that the computed value for helium-building came out lower than the observed value, *i. e.*, $\mu = 0.30$ instead of $\mu = 0.35$, at first introduced a difficulty, since for theoretical reasons our observed value ought to have been lower rather than higher than the theoretical. This difficulty has now been rectified by Klein and Nishina's correction of Dirac's formula which a little more than doubles the theoretical value, and really makes a better fit with our recently repeated observations than we had before.

seem to be inimical at least to the process of the creation of the foregoing abundant elements out of the primordial positive and negative electrons. The building of the radioactive elements, which is an endothermic, or energy absorbing, process, may possibly be taking place in the stars where surplus energy is available for it. We have no experimental evidence whatever on this point. But we have what we consider excellent experimental proof that the foregoing endothermic processes that produce the cosmic rays do not take place in the stars at all. The full argument is given in the October issue of *The Physical Review,* 1928, and in another article to appear in 1930 in the same journal, but the fact that the sun, the great hot mass just "off our bows," has no influence whatever upon the intensity of the observed cosmic rays, for these come in just as strong at midnight as at noon, is enough to show that this particular star is not a source of cosmic rays. Since, however, these rays do come in to us all the time, and practically uniformly from all directions, Dr. Cameron and I can find no escape from the conclusion that these atom-building processes which give rise to the observed cosmic rays are favored by, and actually have their source in, the places in the universe where the temperatures and pressures are extreme in the opposite sense, *i. e.*, where they are close to absolute zero. In other words, *we think that the atom-building processes that give rise to the observed cosmic rays can take place only under the extreme conditions of temperature and pressure existing in interstellar, or intergalactic, space.*

Now combine this conclusion with that already arrived at by astronomers like Eddington and Jeans, who can find no way of accounting for the immense quantities of energy which for billions of years have been poured out by the sun and other stars, save in the assumption that under the conditions of stupendous temperatures and pressures existing at or near their centres mass is being wholly converted into radiant energy by the complete falling together of positive and negative electrons. If this is a correct conception, and it has become orthodox astronomy, then the combination of it with the cosmic-ray evidence herewith presented *leads to the picture of a continuous atom-destroying process*

taking place under the extreme conditions exist-
ing in the interior of stars, and an atom-creating
process continually taking place under the
equally extreme conditions of just the opposite
sort existing in interstellar space. Let us analyze a little further these two processes.

The process in which positive and negative
electrons under the influence of the stupendous
temperatures and pressures existing in the in-
teriors of stars completely fall together—this
need happen only occasionally in the interior of
very heavy atoms, heavier, Jeans thinks, than
any existing on the earth—must in any case, as
we find from Einstein's equation, be one that
gives rise to an ether-wave about four times as
energetic as the cosmic ray due to the formation
of silicon out of hydrogen. *We have looked*
diligently for a cosmic ray of this sort, but it
definitely does not appear in our cosmic-ray
curves. This is, however, to be expected, since,
according to its sponsors, it is formed only in the
interior of stars, and hence is hidden away behind an impenetrable screen of matter that completely transforms it into heat before it gets out.
Indeed, in accordance with the Jeans-Eddington

theory this is merely the way the furnaces of the stars are continually being stoked and all that we ought to observe is the heat and light that they radiate in consequence.

The continuous formation, however, of the common elements in interstellar space newly and directly observed, as we think, through the cosmic rays thereby sent forth raises imperiously the question as to why the primordial positive and negative electrons, which are the building stones of these common atoms, have not long ago been used up, since the process has undoubtedly been going on for eons upon eons. And the answer that Dr. MacMillan, of the University of Chicago, would like to have us make is that out in the depths of space, where we actually observe, through the cosmic rays, helium, oxygen and silicon being continually formed out of positive and negative electrons, there too these positive and negative electrons are also being continually replenished through the conversion back into them, under the conditions of zero temperatures and densities existing there, of the radiation continually pouring out into space from the stars. With the aid of this assumption one would be able to regard the universe as in a steady state now, and also to banish forever the nihilistic doctrine of its ultimate "heat-death."

We ourselves regard this assumption as the least radical and the most satisfactory of any of the three between which, in any case, a choice must be made.

1. The first of these is that of Jeans,[1] that mass, *i. e.*, the electron (positive and negative alike), is convertible into radiant energy, *but that the process is nowhere reversible.* Matter will thus ultimately be all converted into radiation, *i. e.*, it will simply disappear. A recent statement of Jeans' reads: "Thus observation and theory agree in indicating that the universe is melting away into radiation. Our position is that of polar bears on an iceberg that has broken loose from its ice pack surrounding the pole, and is inexorably melting away as the iceberg drifts to warmer latitudes and ultimate extinction."

This is the old hypothesis of the "heat-death." It conflicts with no observed facts, and before the advent of Einstein it was a necessary

[1] J. H. Jeans, *Nature*, *121*, 467, 1928.

consequence of the Second Law *provided the universe were treated as a closed system.* Scientists, however, have always objected that such treatment represents an extravagant and illegitimate extrapolation from our very limited mundane experience and modern philosophers and theologians have also objected on the ground that it overthrows the doctrine of Immanence and requires a return to the middle-age assumption of a *Deus ex machina.* Since the advent of Einstein it meets the further difficulty that it injects into modern thermodynamics one single process—the convertibility of mass into radiant energy—which violates the principle of "microscopic reversibility" required by the modern statement of the Second Law.

2. The second possible hypothesis is that of Stern,[1] Tolman[2] and Zwicky,[3] that the foregoing processes are all everywhere reversible. This hypothesis keeps the second law intact, including microscopic reversibility denied by Jeans' assumption, but so far as we can now see

[1]O. Stern, *Zeit. f. Elektrochemie, 31,* 448, 1925.
[2]Richard C. Tolman, *Proc. Nat. Acad. Sci., 12,* 67, 1926; *14,* 268, 348, 353, 1928.
[3]F. Zwicky, *Proc. Nat. Acad. Sci., 14,* 592–597, 1928.

it does not avoid the "heat-death," and it is not favored by the evidence herewith presented that the atom-building processes that give rise to the cosmic rays do not seem to be taking place everywhere, *e. g.*, in the stars, but do seem to be taking place solely in the depths of space.

3. The third hypothesis—that herewith presented—is just as radical as 1, but no more so, in denying microscopic reversibility, but it provides an escape sought in vain by both 1 and 2 from the "heat-death." Also it is just as radical as 2, but no more so, in assuming that radiant energy can condense into atoms somewhere, but it is in better accord with the cosmic-ray evidence that the atom-creating processes seem to take place only in interstellar space.

But if the point of view developed in the foregoing is correct what sources of energy are there, then, for man to draw upon during the next billion years of his existence? The answer has already been given but it may be restated thus:

(1) The energy available to him through the *disintegration* of radioactive, or any other, atoms may perhaps be sufficient to keep the

corner peanut and popcorn man going, on a few street corners in our larger towns, for a long time yet to come, but that is all.

(2) The energy available to him through the *building-up* of the common elements out of the enormous quantities of hydrogen existing in the waters of the earth would be practically unlimited provided such atom-building processes could be made to take place on the earth. But the indications of the cosmic rays are that these atom-building processes can take place only under the conditions of temperature and pressure existing in interstellar space. Hence there is not even a remote likelihood that man can ever tap this source of energy at all. The hydrogen of the oceans is not likely to ever be converted by man into helium, oxygen, silicon or iron.

(3) The energy supplied to man in the past has been obtained wholly from the sun, and a billion years hence he will still, I think, be supplying all his needs for light, and warmth, and power entirely from the sun. How best to utilize solar energy it is not the purpose of this paper to reveal. That subject is treated in mas-

terly fashion in a paper by Edwin E. Slosson entitled "The Coming of the New Age of Coal," printed in the *Proceedings* of the International Conference on Bituminous Coal held from November 15 to 18, 1926—a paper to which I refer the reader for the next chapter on "Available Energy." The present paper serves merely as an introduction to Dr. Slosson's.

(4) When the matter of the sun has all been stoked into his furnaces and they are gone altogether out another sun will probably have been formed, so that on this earth or on some other earth—it matters not which some billion of years hence—the development of man may still be going on.

V

THE LAST FIFTEEN YEARS OF PHYSICS[1]

WITH full recognition of the tendency of each generation to consider itself important and its contributions to progress unique, I feel altogether confident that the historian of the future will estimate the past thirty years as the most extraordinary in the history of the world up to the present in the number and the fundamental character of the discoveries in physics to which it has given birth, and in the changes brought about by these discoveries in man's conception as to the nature of the physical world in which he lives.

There has been no period at all comparable with it unless it be the period about 300 years ago, which saw the development of Galilean and Newtonian mechanics. This was indeed of incalculable importance for the destinies of the race. The conceptions then introduced are not only the basis of modern *material* civilization, but they were the cause of a very complete

[1]A paper read before the American Philosophical Society in the spring of 1926.

change in man's whole intellectual and spiritual outlook—in his philosophy, his religion, and his morals. But the discoveries in physics of the past thirty years justify the expectation, at least, of as great if not greater consequences, though of a somewhat different sort,—consequences, too, which are already beginning to be realized.

To appreciate how stupendous a change these discoveries have already wrought in human thought, it is only necessary to reflect that of the six basic principles which at the end of the nineteenth century acted as the police officers to keep the physical world running in orderly fashion, namely:

1. The principle of the conservation of the chemical elements,
2. The principle of the conservation of mass,
3. The principle of the conservation of energy,
4. The principle of the conservation of momentum,
5. The principle underlying Maxwell's electrodynamics,
6. The principle of entrophy or the second law of thermodynamics,

there is not one the universal validity of which has not been questioned recently by competent physicists, while most of them have been definitely proved to be subject to exceptions.

The principle of the conservation of the chemical elements went with the discovery of radioactivity. The principle of the conservation of mass vanished with the experimental discovery of the increase in the mass of the electron as its speed approaches the velocity of light. The principle of the conservation of energy suffered at least a change in aspect when both experimental and theoretical evidence came forward that energy and mass are interconvertible terms related by the Einstein equation $MC^2 = E$, for in that equation the ideas of energy and mass become completely scrambled. The principle of the conservation of momentum is denied *universality* by the quantum theory. The Maxwell equations are violated in atomic mechanics. The principle of entrophy was admitted to have exceptions as soon as entrophy was interpreted in terms of probability.

Further, the speed with which new discoveries and new points of view are coming into modern physics shows as yet no abatement. Fifteen years ago it was thought that the revolution had pretty well spent its force, that the main group of new ideas had already been introduced. But a listing of the most outstanding discoveries of the past thirty years shows that the great majority of them belong exclusively to the period here in review—namely, the last fifteen years—and all of them belong at least in part to this period. The immensity of the progress made within it will be best appreciated through a rapid review of this whole list and brief comments upon the origin and the particular significance of each discovery.

(1) *The discovery of the Electron.* This was a very gradual process covering about 150 years and participated in by many workers, Franklin, Faraday, Weber, Helmholtz, Stoney, Lorentz, Zeeman, J. J. Thomson, Lenard, Townsend, Wilson and others,[1] but the actual isolation and the exact measurement of the electron occupied the first seven years of the period in review, 1910–25.

[1] For the historical review see "The Electron," Univ. of Chicago Press, chapters I–V, 1924.

(2) *The discovery of X-rays.* This falls clearly outside the present fifteen-year period, but practically the whole of the *quantitative working* out of the properties of X-rays and the great discovery in 1912 of their wave nature[1] lies wholly within it.

(3) *The discovery of Quantum Mechanics.* This began at about the year 1900 with the work of Planck, but the last fifteen years have contributed enormously to it, as will appear as the enumeration of discoveries proceeds.

(4) *The discovery of the Principle of Relativity.* Though the special principle dates from about 1905 and therefore lies within the first half of the last thirty-year period, the formulation by Einstein of the general principle belongs wholly to the period in review. Its birthday was in 1915, and its most precise and significant experimental verification through the measurement of the bending of starlight in going past the rim of the sun, the discovery of the enormous spectral shift of lines coming from

---

[1]Von Lane, Friedrich, and Knipping Sitz, "Ber. der München Akad.," 1912; also *Jahrbuch für Radioaktivität u. Elektronik*, II, 303, 1914.

the companion of Sirius,[1] and the beautifully consistent measurements of the displacements of solar spectral lines,[2] has all been the work of the past few years.

(5) *The discovery of Radioactivity.* This belongs indeed wholly to the first half of the thirty-year period, but the definite proof that lead is a product of the radioactive disintegration of both uranium and thorium, and the application of this fact to the fixing of the minimum age of certain uraninites from the Black Hills in Dakota as 1,677,000 years,[3] to take but a single concrete case, is very new and marks an important advance in the process of elaborating a very much more definite geologic time scale than has heretofore been available.

(6) *The discovery of the Nuclear Atom* through the experiments on *a*-ray scattering, begun about 1912 and carried on for ten years, mostly at Manchester and at Cambridge, England,[4] falls wholly within the last fifteen-year

[1]Walter A. Adams, Mt. Wilson Observatory Contributions, 1925.
[2]Charles E. St. John, Mt. Wilson Observatory Contributions, 1925.
[3]C. W. Davis, *Amer. Jr. Science*, March, 1925.
[4]Rutherford, Geiger, Marsden, and Chadwick, *Phil. Mag., 21*, 699, 1911; *25*, 604, 1913; *40*, 724, 1920; *42*, 933, 1922.

period and has been epoch-making in its consequences.

(7) *The discovery of Crystal Structures* through the aid of X-ray spectroscopy[1] dates from only 1913. It has completely revolutionized crystallography and opened up a new world of definite knowledge about molecular and atomic arrangements in solids.

(8) *The discovery of Atomic Numbers* (1913–1924) and the definite fixing of the total number of possible elements between hydrogen and uranium as 92,[2] both being included, is perhaps the most beautiful and the most simplifying discovery ever made. Nature never came so near surrendering herself to her lover without reserve and revealing herself in beautiful and simple grandeur as when Moseley found that all the elements fitted into a single arithmetical progression—a progression, too, which could mean nothing except that the positive charge on the nucleus of each atom moved up by unit steps from 1 to 92 in going from hy-

[1] Bragg, "X-rays and Crystal Structure," Bell, London, 1916; see also Ewald, "Kristalle u. Röntgenstrahlen," Springer, Berlin, 1923.

[2] Moseley, *Phil. Mag.*, 26, 1024, 1913; 27, 703, 1914.

drogen to uranium—a progression which at once robbed atomic weights of their long-usurped right to act as arbiters of the chemical destinies of atoms, and restored this place to its legitimate possessor, the electrical charge of the nucleus. Some day a poet will arise who will make an epic for the ages out of young Moseley's discovery. The X-ray spectroscopists[1] and those who have recently extended the X-ray laws into the field of optics,[2] have contributed to the establishment of the complete generality of the Moseley progression, but they have but finished the structure which he designed.

(9) *The discovery that the Energy communicated to electrons by ether waves is proportional to the frequency of the absorbed waves.* This was vaguely suggested by Planck in 1900, more specifically by Einstein in 1905, but the experimental proof of its correctness is perhaps the most momentous achievement of modern physics, and it has all come about since 1912. The achievement is momentous not merely because the equation $\frac{1}{2} mv^2 = h\sqrt{} - P$

[1]Siegbahn, "Spektroskopic der Röntgenstrahlen," Springer, Berlin, 1923.
[2]Millikan and Bowen, *Physical Review*, 1926.

ranks with the equations of Maxwell in its consequences, but because the relation itself is new,
undreamed of in 1895, and altogether revolu-
tionary, demanding a return to some elements
of the corpuscular theory of ether waves. It
was proved first very exactly in photoelectric
experiments with light waves,[1] then during
and after the war with X-rays,[2] and then with
γ-rays. Most of the preceding discoveries were
wonderful *additions* to knowledge, explora-
tions in heretofore unknown fields but not
subversive of established conceptions. This one,
however, wrought havoc with existing theories
and demanded a new formulation of ideas about
the relations of ether physics and matter physics. Its significance for the future can scarcely
be overstated.

(10) *The discovery of the meaning of Spectral Lines.* Bohr[3] in setting up his theory of
atomic structure merely generalized the preceding relation. With unusual insight into the
method of constructive science, he incorporated

[1]"The Electron," chapter X.
[2]Duane, *Physical Review*, 7, 599, 1916; *9*, 568, 1916; *10*, 93 and
624, 1917; de Broglie, Third Solvey Congress, 1921; Ellis, *Proc.
Roy. Soc.*, *99*, 261, 1921; also same, Jan., 1924.
[3]N. Bohr, *Phil. Mag.*, *26*, 1, 476, 857, 1913; see also "The Electron," chapter IX.

all the past and merely superposed upon celestial mechanics the assumption, almost inevitable (if Einstein's equation is correct), that when an ether wave is *emitted* as well as when it is being absorbed, the foregoing relation between energy and frequency still holds, *i. e.*, that the emitted wave frequency is given by $E_2 - E_1 = h\sqrt{}$, the E_2 and E_1 being the electronic energy before and after emission. Combining this with the experimentally established Ritz-Balmer equation, he brought out sharply the unitary or atomic character of angular momentum, clearly one of the two or three most fundamental discoveries of all time, for it will presumably always be the basis of all atomic mechanics. It had been dimly glimpsed before in the work of Planck and Einstein, definitely stated by Nicholson and Ehrenfest, but from Bohr's time on quantum theory took on a definiteness, almost a visualizability before unknown.

(11) *The discovery of isotopes.* This discovery did not begin to be made until 1913, when chemists and physicists approached it from two different angles; first, the chemistry of the

radioactive elements, and, second, positive ray analysis.[1] It was completed only aften ten years of work by physicists in analyzing by positive ray methods forty-six of the first fifty-five elements of the periodic table. Its significance lies in the four following facts: first, that it has resurrected completely the discredited and amazingly simple Prout hypothesis that the masses of all atoms are exact multiples of the mass of a primordial atom; second, that it has enabled us to count with certainty the exact number of positive and negative electrons inside every nucleus—positives being equal in number to the atomic weight, negatives to the atomic weight minus the atomic number; third, that the failure of hydrogen to fit exactly into the above scheme constitutes excellent evidence for the Einstein conclusion as to the interconvertibility of mass and energy; and fourth, that the difference in stability (decay-time) of radioactive isotopes suggests new possibilities in the reading of the structure of the nucleus, *i. e.*, it gives us new eyes for peering inside the tiniest organism yet found—the nucleus of the atom.

[1] See Aston's "Isotopes," London, 1922.

(12) *The discovery of "the excited atom."*
The purely theoretical reflections in 1921 of
two mere youths in Denmark, Klein and Rosse-
land, resulting in the proof that if atoms can be
thrown by impact, as experiment shows that
they can, into a quantum state of higher energy
than the normal,—appropriately called an excited state—then unless the second law of thermodynamics is to be violated, there must be a
heretofore unrecognized mechanism by which
an excited atom can return to its normal state
without radiating at all, but rather by throwing
back all the potential energy of its excited condition through a so-called "collision of the
second kind" into an electron or an atom projected from the collision with a kinetic energy
that may be 100 times the average energy of
molecular agitation. This is a very recent discovery of the first magnitude and of possibly
immeasurable significance. It has already made
it possible through the work of the experimental
physicist[1] to see a collision-mechanism by which
a negligible number of mercury atoms can ab-

[1]See Loria, *Phys. Rev.*, Nov., 1925, and *Proc. Nat. Acad.*,
Dec., 1925, for review of experimental work of Franck, Caria,
Donat, and Loria. See also Franck's recent book.

sorb ether-wave radiations and transfer that energy to the act of exciting thallium atoms to radiate their characteristic frequencies. It for the first time reveals a definite mechanism, sought in vain by the best thinkers of the nineteenth century, by which the energy of ether waves may be absorbed by matter and transformed into heat. It takes in one case, at least, the mystery out of the words "catalytic agent." It suggests why very minute quantities of vitamines, etc., may be the intermediaries through which very vital processes are brought about. It makes the future bright with promise for the better understanding of explosive processes. Indeed, the properties of excited atoms may be the foundations of a new era in both industrial and biological sciences.

(13) *The discovery of the artificial disintegrability of atoms.* This was glimpsed about 1912 through the appearance of hydrogen lines where hydrogen, unless artificially produced from other elements, should not have been present, and new claims of spectroscopic evidence of a similar sort are now being advanced; but the unambiguous proof is contained in Ruther-

ford's[1] direct experiments showing that hydrogen atoms can be knocked out of other atoms by alpha-ray bombardment. The method shares with that which will introduce new resolution into the study of the masses of isotopes, the promise for the future reading of the conditions of the electrons within the nucleus.

(14) *The discovery of the meaning of spectroscopic fine structure.* It was as late as 1917 before Sommerfeld[2] began his wonderful work on the interpretation of the fine structure of spectral lines—work which began for the first time to reveal the correct principles underlying quantization, finally so penetratingly formulated by Epstein. Seldom in the history of physics have purely theoretical formulæ had such amazing successes in the field of precise prediction as have Sommerfeld's relativity-doublet formulæ and Epstein's extension of the same sort of orbit considerations to the prediction of the number and character of the multiplicity of lines found in the Stark effect. The whole interpretation of spectroscopic fine struc-

---

[1]Rutherford and pupils, *Phil. Mag.*, 1920–1925.
[2]Sommerfeld, "Atombau u. Spectrallinien," Viewegn **Sohn**, **Braunschweig**, 1924, chapter VI.

ture through changes in so-called azimuthal and inner quantum numbers is one of the great achievements of all time resulting from the interplay between penetrating theoretical analysis and skilful and refined experimental technic.

(15) *The discovery of New Experimental Technics for seeing Invisible Ether Waves.* Such long-wave technic[1] has completely bridged within the past two years the gap between artificial electromagnetic waves and heat waves, and in the short-wave region hot-spark vacuum spectrometry[2] and beta ray methods of analysis have practically filled in completely the gap between the optical and the X-ray fields, while far above even the gamma rays of radium a new group of rays of well-nigh infinitely high frequency has been found. (See below.) This continuous passage of frequencies from a thousand billion billion per second, over into the zero frequency, *i. e.*, over into static electrical fields, all these waves possessing identical characteristics as to speed of propagation, as to polarization, and as to relations of electric and

[1]Nichols and Tear, *Phys. Rev.*, 1925.
[2]Millikan, *Astro. Phy. Jr.* for 1920, and Millikan and Bowen, *Phys. Rev.*, 1923 and 1925.

magnetic vectors, demands one and the same sort of transmitting mechanism or medium to take care of them all, by whatever name it may be called, whether a "world-ether" or "space," this last term meaning no longer emptiness, but emptiness endowed with definite properties, if such a Hibernianism suits one's taste.

(16) *The discovery of new properties in Conduction Electrons.* The direct measurement[1] of the mass of conduction electrons, the discovery of a general increase in conductivity with the application of enormous pressures,[2] of superconductivities,[3] and the very recent proof that, when intense electric fields pull conduction electrons out of metals, the energy of thermal-agitation assists not at all at ordinary temperatures, but does assist at very high temperatures, all these recent discoveries begin to clear up the contradictions in the electron theory of metallic conduction, and to enable the quantum theory of specific heats to be applied in a new and illuminating way to the condition of the

[1]Tolman, R. C., *Phys. Rev.*, 1920–1929.
[2]Bridgman, *Phys. Rev.*, 1925.
[3]Kammerlingh Onnes, Contributions from the Univ. of Leiden, 1910–1925; Millikan and Eyring, *Phys. Rev.*, Jan., 1926.

electrons in metals.[1] The new results are full of promise for the better understanding in the near future of the moot subject of metallic conduction.

(17) *The discovery of Quantum Jumps inside the Nucleus.* The very recent proof[2] that gamma radiations obey the same quantum jump laws obeyed by X-rays and light rays and the still more recent proof that the initial act in a radioactive change is the ejection of an alpha or beta ray—possibly by virtue of the actual loss of mass of the nucleus through electronic settling and the transformation of this mass into the energy of the ray, following the Einstein relation governing the interconvertibility of mass and energy—this is a discovery of importance for the future understanding of one of the most fundamental processes in nature, the growth and transmutation of the elements.

(18) *The discovery that the law of the Conservation of Momentum is applicable to the encounter between a light quant and a free electron.* This very recent and very amazing dis-

---

[1] Sommerfeld's and Houston's recent papers, 1928 and 1929, on metallic conduction seem to mark important advances.

[2] Ellis and Rutherford, *Proc. Roy. Soc.*, 1925 and 1926; Meitner, *Zeit. f. Phys.*, 1925.

covery[1] seems to put the final nail into the Einstein conception of radiant energy travelling through space in the form of vibratory light-darts of some sort but at the same time it emphasizes the apparent impossibility of the physicist's finding in his present stage of development any one consistent and universally applicable scheme of interpretation. It is a discovery of the very first magnitude, one of whose chief values may be to keep the physicist modest and undogmatic, still willing, unlike some scientists and many philosophers, not to take himself too seriously and to recognize that he does not yet know much about ultimate realities.

(19) *The discovery of the Summation of two or more Quantum Jumps into a single monochromatic ether wave.* The very recent proof brought forth first by the properties of band spectra[2] and second by the discovery of two electron jumps through study of pp′ groups in line spectra,[3] that an atom can integrate the combined energy of two distinct and simultane-

[1] A. N. Compton, *Phys. Rev.*, 1923, 1924, and 1925.
[2] Sommerfeld, "Atombau u. Spectrallinien," chapter IX.
[3] Russel and Saunders, *Astro. Phys.* of 1925; Wentzel, *Zeit. f. Phys.*, 1925; Bowen and Millikan, *Phys. Rev.*, 1925.

ous quantum jumps into a single emitted mono-chromatic wave is of fundamental importance, because of the new light which it throws upon the nature of the act in which a ray of light is born and projected on its way through space. The discovery of two electron jumps uniting into a single monochromatic ether wave seems to preclude the possibility that there is any vibrating mechanism in the atom executing vibrations synchronously with the period of the emitted monochromatic wave. That the atom has this power of transforming, in some as yet mysterious way, the energy of every atomic shudder into a monochromatic ether wave is amazing. Whether we shall ever be able to visualize the process more definitely than we can now no one knows.

(20) *The discovery of the failure of the Relativity Explanation of all Relativity-doublets.* The impasse pointed out within a year[1] between the heretofore recognized causes of doublets in optics and in X-rays, an impasse which necessitated the finding of a new cause which would yield exactly the same formula as

[1]Millikan and Bowen, *Phil. Mag.*, 1925. Also Landé, *Zeit. der Phys.*, 1925.

the relativity cause—a really terrible necessity in view of the extraordinary quantitative success of the purely theoretical formula following from the mere postulation of the change of the mass of the electron with speed—this impasse has apparently just been resolved within two months by two young Dutchmen, Oulenbech and Goudsmit, who find, "mirabile dictu," that the assumption that every electron in the universe spins with one unit of angular momentum either right-handedly or left-handedly yields a formula of exactly the same form as does the relativity cause. This assumption not only saves the relativity explanation but combined with it furnishes a much better correlation of all present spectroscopic facts than we have heretofore had. It represents probably a fundamental advance in our understanding of the nature of the most nearly ultimate thing with which physics deals, namely, the electron. The physics of the future bids fair to hear much of the *spinning electron.*

(21) *The discovery of Cosmic rays.*[1] That

[1] For a review of the literature see "High Frequency Cosmic Rays," *Phys. Rev.*, May, June, and July, 1926. See also Smithsonian Reports for 1929.

something very fundamental is going on all through space, that nuclear transformations, each of enormous energy value, corresponding to the fall of an electron through as much as 200,000,000 volts, are actually taking place in all directions about us in the outer stretches of the universe and that the signals of these cosmic changes can be detected here has recently been proved. It is a discovery most stimulating to the imagination. Call it the music of the spheres if you wish! Anyway man can hear it now and may some time know more about it.

Of these twenty-one fundamental discoveries at least sixteen fall wholly within the past fifteen-year period, and all of them have belonged in no small degree to it. Physics as yet seems to show no signs of approaching senility. What the future has in store, no man knows, but at present there seem to be well-nigh limitless possibilities ahead for application even if the pace of discovery should some time slacken.

VI

THE FUTURE OF STEEL[1]

FIFTY thousand years ago, or perhaps five hundred thousand, when the progenitor of modern man first got through his thickened skull the idea that he could more safely vanquish his antagonist or strike down his quarry by hurling something at his head than by sinking his teeth into his throat, that emerging man of course picked up the most abundant material about him for his weapon and the "stone age" was born.

How long did it last? For scores if not for hundreds of millenniums, so slowly does man's knowledge of what is about him in his physical world advance. It had not entirely disappeared when David with his sling sank a smooth stone from the brook into the forehead of Goliath.

But despite their abundance, stones make poor tools. They are too brittle. A blow shatters them. They are too difficult to shape and above all to sharpen. One day, perhaps ten

[1]An address delivered before the American Iron and Steel Institute in New York in May, 1929.

[135]

thousand years ago, some stone-age man picked up, or dug up, a piece of native copper. It was four times as heavy as his stone and therefore better adapted to break an opponent's head. It could not possibly be splintered, but could be pounded into new and convenient shapes, especially when it had been heated very hot in the fire, the use of which from the remotest times and among all peoples of whatever stage of development has distinguished mankind from all the rest of creation.

With this discovery the maker of flint arrowheads and stone axes lost his job. A new industry was born immensely more varied and more satisfying than that which it replaced.

And let it be noted here incidentally that this is always what happens with every advance in man's knowledge of his world. Let him whose eyes have recently been focussed on the increased effectiveness of tools of destruction, and whose fears have been aroused lest the savage in man may use these tools to destroy the race, take off his blinders and look around him on all sides. I think such a survey will show him conclusively that *every scientific advance finds ten*

*times as many new peaceful, constructive uses as
it finds destructive ones.*

Let him who would slow down the march of
science, which is simply the march of knowledge, of understanding of our world, reflect on
this. When the age of metals replaced the age
of stone the arrowhead and the tomahawk
maker did indeed lose their jobs, but new and
altogether wonderful peaceful arts were born.
The metal workers of the bronze age with all
the decorative art that they developed for use
on urns and vases and household wares, and
sarcophagi and friezes—an art which excites our
deepest wonder and admiration to this day—
these men developed new capacities of expression and of appreciation in the race. Civilization shot forward through their work.

But the bronze age too was doomed, though
not the art that developed with it, nor even the
use of bronze itself. And note again that this
is what usually happens, though not always,
with the advance of knowledge. Every new
creation in art or science is from the day of its
birth the heritage of all succeeding ages, and if
it disappears it will usually be because it doesn't

represent a real advance, *i. e.*, because it is pseudoscience or pseudoart. Let me hasten to admit, however, that this is not always true. The breeding of draft horses as an industry has undoubtedly been ruined by the advent of the automobile. Whether this represents a gain or a loss to horse-civilization is a debatable question. It may be that horse-well-being and the mean of horse-happiness has been increased thereby. One's judgment on this point will depend upon the particular kind of horse-sense that one uses in the analysis.

But be that as it may the passing of the bronze age certainly did not represent the loss of the metal arts. The mining and working of other metals had followed or accompanied that of copper. Artificers in gold and silver and zinc and lead had appeared and new uses for all these metals had been found. The age of metals had forever replaced the age of stone.

But one of these new metals which came late into the field usurped the throne completely and ushered in the age of iron, the age in which we live. It has been with us for perhaps three or four thousand years, a minute fraction of the age of stone, shorter than the age of bronze; but it is only within the past hundred years, or since the advent of the industrial revolution, that iron and steel have assumed so altogether outstanding, so supreme a place in human life. And we are gathered here to-night at the meeting of the American Iron and Steel Institute in 1929 to ask when it, in its turn, must pass, and give place to something as much more potent than it as it was more potent than bronze, or bronze more potent than stone.

I think the answer can be given, but let me ask, first, how much is it worth, to an industry that presumably has an annual payroll of as much as a billion dollars, to know it, and let me ask, second, whence has come this knowledge that should be so vital to this industry?

The first question I leave you to answer, but I suspect that, in this rapidly changing world, this world in which new methods make expensive equipment archaic over night, this world in which with the rapid extension of mass production every change becomes more and more difficult and more and more costly, I suspect that any assurance of stability, or even any foreknowledge as to the direction in which changes are likely to be demanded, is worth in stock market quotations alone, millions upon millions of dollars, to an industry like that of iron and steel. I leave you to estimate how many millions.

But as to the second question, whence is that knowledge to come which makes any sort of prediction possible? The answer to that is certain. It is not from anything whatever that you have done or are doing in your plants and factories, not in the main from any work to which you have directly or indirectly contributed as a steel industry, but rather *from fundamental work in the field of pure science,* work on the constitution of matter that has been going on chiefly in university laboratories supported in the main merely by funds contributed by far-seeing philanthropists, and by the devotion of men who have asked practically nothing for themselves except the chance to help to further the progress of humanity, the chance to add something to our knowledge of what nature's laws are, and hence of how we can learn to live more wisely in a world governed by them.

The knowledge already acquired about the constitution of matter, and which I estimate is already worth millions upon millions to your industry is very new. Eighteen years ago we did not know how many elements there were on our earth, nor when some one might discover a new one that might do to iron just what iron had done to copper and copper to silicon.

To-day we know, thanks to developments in pure science just how many elements there are on earth, and approximately what the properties of all of them are, and in view of that knowledge *we cannot see any metal that is likely ever to compete with iron, for structural purposes, for meeting the big demands of transportation, or for satisfying the chief requirements of power machines.*

But cheapness, availability, abundance is no less important than physical properties. Where must you go to get a reliable estimate as to the *abundance* of the various elements? Again not to the steel industry but to the pure research laboratory. Oliver C. Farrington of the Field Columbian Museum has spent the greater part of his life on determining the constitution of meteorites, a problem that no industrial laboratory would ever dream of tackling. The results are interesting and important. By weight 95 per cent of all meteorites consist of the four elements, oxygen (54 per cent), magnesium (13 per cent), silicon (15 per cent) and iron (13 per cent). This is perhaps as near as we can come to estimates *of the relative abundance of the solid elements in creation.*

To get checks upon these estimates, however, we can turn to two other sources; first to the spectroscopy of the heavens, carried on wholly by astronomers, the group whose field is usually regarded as the most useless of all the fields of science; and second to surveys of the constitution of the earth's crust. The best of these surveys has also been compiled by an astronomer, using data I think wholly obtained from pure research laboratories. These three methods of approach lead to the same general conclusion.

Thus 76 per cent of the earth's crust is composed of the three elements, oxygen (55 per cent), silicon (16 per cent), aluminum (5 per cent), no other element rising over 2 per cent.

Iron constitutes 1.5 per cent of the crust but is supposed to be very largely represented in the interior.

The evidence from the spectroscopy of the stars is less definite but it too assigns a very great abundance to the foregoing elements and gives no others a prominent place unless they be calcium and sodium. Calcium constitutes 1.5 per cent of the earth's crust, while sodium is 2 per cent of the earth's crust and is negligible in meteorites.

Silicon then appears to be the next most abundant element to oxygen, the only others that need be considered at all in a survey of all the elements being magnesium, aluminum and iron. Silicon has already had its contest with iron, and there is no danger now that we shall ever return to the stone age, though stone will doubtless always retain something like the place that it holds now.

Among the metals there is no even remote competitor with iron from the standpoint of abundance except magnesium and aluminum. If lightness were a desideratum these two might possibly compete for the throne of iron. But for the great bulk of the purposes for which iron is used, for all kinds of structural purposes, for rails, for stationary power machines, even for automobiles, lightness is not desirable; so that while our greatest industry of to-day, agriculture, may conceivably some day shrink to small proportions because of the development of synthetic foods, while another great industry, textiles, may lose half of its present market because the female half of the population prefers to go unclad—the curve of female apparel is just now moving very rapidly toward that asymptote—yet the iron and steel industry can sit and watch the passing show with entire complacence and assurance that millions of years hence as now the world will give it what it wants—its trade. The future of steel is assured!

But the future of the industries that constitute the American Iron and Steel Institute is not assured, unless we in America do our full share in developing the foundations on which those industries rest.

What are those foundations? The answer has already been given, but a brief review of the history of steel will make the road so clear that the wayfaring man cannot miss it.

Iron gained its first pre-eminence over the other metals through the discovery of the extraordinary qualities that a small admixture of carbon imparted to it, and most, if not all of succeeding advances in the discovery of new qualities and new uses have been due to the finding of new alloys. What alloys? Their names tell the story, nickel-steel, chrome-steel, vanadium-steel, uranium-steel, etc. But what made these alloys possible? *The prior discovery in the pure science laboratory of the new element used in each alloy, in other words fundamental pure-science-studies on the constitution of matter have underlain all your advances.*

But the study of alloys themselves has up to just this moment been a hit-or-miss affair. Through fundamental studies in the field of X-rays, pure science is just now in possession of a rational and systematic mode of approach to the problem of alloys, and in the course of time results of the greatest importance such as cheaper, stronger, more durable materials for all purposes, are almost sure to come.

[145]

Where are these fundamental advances most likely to be made? *Obviously in the future as in the past in the pure science laboratory.* Your industrial laboratories must be awake to them, ready to make the applications as they appear, but underneath it all, as a basis for it all, unquestionably lies the field of pure science.

But, I hear some one inquire, why cannot our industrial laboratories in the United States keep watch over the developments in pure science all over the world and make the applications as soon as they appear? What especial need is there for supporting pure science in this country? The answer is *"If we are not developing in this country the brains that can make the advances, neither shall we have the brains that can make the applications."* Our industries are recruited in their more technical branches from our universities. As industry becomes more and more complex, such recruiting must take place in ever-increasing amount. Keep the pure science keenly alive in the universities and the applied science laboratories of the industries will draw from them, catch of their spirit and participate in their activity. But once let the fountain dry up at the source and the whole valley through which it ought to flow will become a desert. Take it as certain that if America cannot lead in science neither in the long run can she lead in industry. The two call for precisely the same sort of brains and the same sort of energies.

No country has ever had such an opportunity as ours, such a wide-spread stimulation of individual initiative, such a large number of men and of groups who have learned to treat financial and business power as a public trust, such a widely diffused attitude of co-operative social effort, such ability to utilize it, all because of our individualistic system and our centuries of training in depending upon ourselves instead of upon government, such resources to command, such results to anticipate. The steel industry is now one of the most influential in the country. Its leadership in this matter will be no less influential than it is in other matters. To it I would say as Mordecai said to Esther, "It may be that thou art come to the kingdom for such a time as this."

VII

MICHELSON'S ECONOMIC VALUE[1]

IN the year 1896 Albert A. Michelson took
a new egg into the nest over which he brooded
—or the department on which he sat—at the
University of Chicago, and after an incubation
period of twenty-five years—so long a time did
it take to prove that the egg had ever been fer-
tilized at all—he at last had it hatched and sufficiently feathered to justify pushing it out of
the nest and bidding it go scratch up its own
worms.

To-night, Mr. Chairman, you, representing
the public which is obliged to supply the corn-
meal required to keep both Michelson and
Millikan scratching, have brought us here to exhibit our worms and to let you see whether they
are worth the price paid to get them. And as
you will presently see that leaves me no choice
but to take for the subject of my speech the
length of Mr. Michelson's worm, or the eco-

[1]An address delivered in New York on the occasion of the presentation to Michelson and Millikan of the gold medal of the Society of Arts and Sciences.

nomic value of Michelson. For if you ask him to explain, in terms that you can understand, the value of his work, I think that you will be told to go to the interior of a star where the temperature is estimated to be 50,000,000° C., or even to a hotter place, if such there be, described by a familiar monosyllable especially beloved by men like Michelson trained for the sea. For Mr. Michelson is wont to say that the sole reason, and the good and sufficient reason, why he spends so much time trying to measure the velocity of light to one part in three hundred thousand is simply that he likes to do it.

But I am going to make bold, now that I have left the nest and am where he can no longer reduce my rations, to contradict him and to tell you, and to tell him, that that is not the sole reason, nor is it the good and sufficient reason. (You see, Mr. Michelson, the young rooster, after the immemorial manner of young roosters, is questioning the old cock's right to do just exactly as he "damn-pleases" in the hen-yard.) To prove my point I have only to call your attention to the fact that if Mr. Michelson had "chosen" to spend his days and his nights sitting on a log pounding it with the butt end of a hatchet he would soon have found himself in a straight-jacket in the nearest institution especially provided by the state for the care of the deranged.

In a word, the reason you, the public, support Michelson's work on ether drift and the speed of light, and the reason he sticks at those measurements like a burr to a collie-dog, is that both you and he *have faith in the worth-while-ness of those measurements*. And the real reason that he tells the reporters that he does it merely because he likes to is that he knows that if the reporter hasn't already "got religion" it can't be pumped into him in two minutes. There are some things that one cannot explain for the next edition of the tabloid newspaper.

But the matter goes still deeper. It is rarely that a scientist himself knows just where his particular increment to the sum total of human knowledge is going to fit into "the whole," or to find its relations to other increments. Galileo was not looking for useful applications when he was rolling his marbles down the inclined plane

to discover the laws of force and motion, nor
Kepler when he tried to understand the orbits
of the planets, nor Newton when he deduced
the law of gravity from observations on the
moon, nor Volta when, merely toying with con-
tact electrical effects, he laid the foundation for
an understanding of electrical potentials; but
all of them undoubtedly *had confidence in the
value of knowledge* in enabling man to live
more wisely in his world, not perhaps so much
in enabling him to raise more cabbages to the
acre, but, much more important than that, *in
preventing him from wasting his time and his
energies in chasing illusions,* in enabling him *to
direct his thinking and his acting toward realities* instead of toward will-o'-the-wisps. It is
only in looking back from our vantage point of
centuries that we see that these men by their researches were actually *ushering in a new civilization.*

Has the time yet come when we can look
back upon the particular type of activities, exceedingly precise physical measurements, which
Mr. Michelson has followed, and say that from
the standpoint of the hard-boiled business man

who has had to pay the bills they have justified themselves? It is much too early now to balance the books, and will probably be too early for centuries to come, for the inspiring thing about work in the field of science is that *every bit of new knowledge becomes from the moment of its discovery the heritage of all future ages,* enabling coming man, as long as mankind endures, to live just so much more wisely than past man has known how to live. But in the case of Michelson's work I think we can even now get an inkling—a mere suggestion—of some of its economic values, or at least its economic possibilities. Of course these may be either positive or negative—there are always some entries on both sides of a ledger—and since Mr. Michelson always inspired me by example to be strictly honest whatever happens, I am going to begin by presenting one of the liabilities instead of the assets. This one has to do with that particular nearly-bad-egg that he had to incubate for twenty-five years before it hatched and matured sufficiently to justify him in pushing it out of the nest.

When in 1917 I made the best determination

I could of the value of the electron I had to use in getting the final quantity the best value then available of the velocity of light. I therefore went to Mr. Michelson and asked him within what limits I might count upon his determination of the velocity of light. He replied, "To one part in ten thousand." So I chose for my computations the value $c = 2.999 \times 10^{10}$ cm. per sec., the nearest value, to four places of accuracy, to his mean, and since my accuracy in the determination of e could not be more than one part in a thousand I thought I should never have to bother about changing e because of anything that might happen to the velocity of light. But last year, unfortunately for me, Mr. Michelson made a new and most accurate determination of c, the velocity of light, and found it one part in 3,000 instead of one part in 10,000 lower than my chosen value. Also, because of two new determinations of the absolute value of the ohm—experiments of quite the Michelsonian type though made by others—I find that my value of e is probably affected because of the change in this last constant by one part in two thousand, and these two changes being in the same direction, I am obliged to change my value of e by about one part in a thousand so that it becomes 4.770×10^{-10} instead of 4.774×10^{-10}.[1] This is such a wound to my pride that I am thinking of trying to obtain heart-balm by instituting a breach-of-promise

[1] The reason I have not heretofore made the foregoing readjustment in my value of e is, first, that it is of no particular significance anyway (see below); and, second, that I have until recently doubted its legitimacy.

In the presentation of the best values of widely used physical constants I have heretofore questioned the wisdom, or even the correctness, of making a differentiation between so-called international units and absolute units before a suitably authorized international commission had recognized that difference, since otherwise such differentiation would rest merely upon some individual's *estimate* of the superior reliability of some particular new determination or determinations over the weighted mean of the whole series of determinations used by the international commission which in 1908 and 1911 fixed upon the international units. However, Professor Raymond T. Birge has called my attention to the fact that in view primarily of the close agreement between new determinations of the absolute value of the ohm by F. E. Smith (*Phil. Trans.*, 1914) and Grüneisen and Giebe (*Annal. Physik.*, 1920), the compilers of tables have actually recently begun to make the foregoing differentiation. It is because of this fact and because of Michelson's undoubted new precision in the measurement of the velocity of light that I have thought it worth while to begin herewith to recognize the effect of these changes upon the value of e.

But this change is actually only of academic interest, since it is in any case within the limits of my estimated uncertainty. The limit of accuracy of the oil-drop method is fixed by the uncertainty in the measurement of the viscosity of air, which enters into it in the $\frac{3}{2}$ power. In my determination of e I estimated the viscosity of air to be known to $\frac{1}{20}$ per cent. If this is an overestimate e would be affected $1\frac{1}{2}$ times as much, so that if Eddington's recent theoretical deduction of e, which is $\frac{1}{2}$ per cent higher than my value, is correct, the viscosity of air must be about .3 per cent higher than I have estimated it to be. I do not think this can be the case.

suit against Mr. Michelson. If I succeed, the economic value of Michelson will be about a hundred thousand dollars less than it is now. So much for the debit side of the account.

Let us now glance at an item or two on the other side of the ledger. The special theory of relativity may be regarded as merely a generalization of the famous Michelson and Morley experiment, another typical Michelsonian attempt to measure with great precision a quantity of fundamental importance, namely, the speed of the earth through the ether. As everybody knows, it came out negative, that is, no such speed, nor any trace of it, could be found, and after forty years of most painstaking repetitions, capped by Michelson last year, it still seems to be impossible to find any speed whatever of the earth with respect to the ether.

Einstein, in 1905, generalized the foregoing result by postulating that it is in the nature of the universe impossible to find the speed of the earth with respect to ether. This postulate rests most conspicuously upon, and historically grew chiefly out of, the negative result of the Michelson-Morley experiment. Now, one of the most important consequences that Einstein drew from the special theory of relativity is that mass and energy are interconvertible terms—that radiant energy, for example, cannot appear without the corresponding disappearance of mass. This startling conclusion, which amounts to the denial of the ancient doctrine of the conservation of mass as distinct from the conservation of energy, has in recent years met with three new and powerful experimental supports. The first is that the assumption that the sun is stoking his own mass into his furnaces, and thereby reducing his waist-band at a rate measured by the scales of 250,000,000 tons a minute, furnishes the only means the astronomer can now find of accounting for the enormous lifetime of the sun and the stars, as attested by both astronomical and geological evidence. The second is that the interconvertibility of mass and energy seems to be an established experimental fact in the special case of an electron moving with a speed close to that of light. The third is that the facts of radiation-pressure, discovered by Lebedew in Russia, and Nichols and Hull in America, about 1901, mean that ethereal radiation has the only distinguishing property of mass, namely, inertia. Further, the quantitative equation of relationship, namely, $mc^2 = E$, in which m is mass in grams, c the velocity of light, and E energy in ergs, follows inevitably from the theory of relativity, and three different quantitative tests of this theory, all involving very painstaking and very precise measurements of the Michelsonian type, have all now yielded results *in quantitative agreement* with the predictions, so that I think that all physicists and astronomers will now agree that the foregoing equation is a safe guide for the theorist of the present and the future. *Historically, it is hard to see how it could ever have been arrived at without both Michelson's own very exact measurements, and others of the kind he has led the way in showing how to make.*

Now, whether that conception has any immediate commercial importance or not, if it is true, as we think it is, it is one of such stupendous significance for man's understanding of the universe in which he lives that its importance transcends all assignable money value, and Michelson's part in bringing it to light makes him

[157]

a bigger world-asset than any billion dollar corporation in the United States, or than all of them put together.

But I am not through yet with the rôle played recently in human progress by refined physical measurements of the Michelsonian type. While Michelson has been driving ether-drift and speed-of-light experiments to the limit of accuracy in America, Aston in England has been setting himself a precisely similar task in determining, by the so-called isotope technic, the atomic weights of the elements.

The first result was the beautiful discovery that the weights of all the elements are exact multiples of a fundamental unit which is close to the weight of the hydrogen atom. I use the word beautiful in describing this discovery because of the amazing simplicity and orderliness with which it endows nature, inevitably suggesting that all the elements have been built out of hydrogen.

But while we are admiring the beauty and simplicity of this generalization, let me call attention to the interesting circumstance that even some of us physicists overlooked for years the result that everybody can now easily see, namely, that the foregoing exact-multiple law, first stated in 1913 and 1914, is in irreconcilable conflict with the preceding generalization, the Einstein equation. For if mass and energy are interconvertible, and if the masses of all atoms are exact multiples of a mass-unit, then there can be no emission or absorption of energy when these units go out of one atom, or into another. In other words, there can be no forces holding atoms together and preventing transmutations, and there can be no evolution of energy when such transmutations spontaneously occur, as they do in radioactive processes. These conclusions are obviously contrary to all experience so there must be something wrong with the isotope law unless we are willing to throw overboard Einstein's equation.

Just what was wrong was brought to light a year ago last summer, for the case of all elements except hydrogen, solely because of Aston's skill in *adding another decimal point to the accuracy with which he could measure the masses of the various atoms*—a typical Michelsonian accomplishment.

[159]

And these new measurements, along with Einstein's equation, enabled Dr. Cameron and myself to get, about a year ago, what seems to me to be the inevitable interpretation of the so-called "cosmic rays"—strange super-gamma rays which bombard the earth incessantly day and night, apparently without any measurable change in intensity with either direction or time. They are best observed and measured by sinking electroscopes in deep, high altitude, snow-fed lakes. Their easily observable ionizing effects are found to decrease rapidly with depth but, according to our as yet unpublished measurements of last summer (1928), not to disappear entirely until depths of much more than three hundred feet have been reached.

With the aid of Einstein's equation and Aston's atomic weights Dr. Cameron and I have computed the energy that should be released in a radioactive change—a process which transforms one atom of accurately measured mass into another atom of accurately measured mass. The difference in the measured masses before and after the change should give, when multiplied by c^2, the energy evolved, also accurately

[160]

known, and *the results are found to check with
the computation,* thus giving credentials to the
method.

The interesting result of this mode of approach is that it shows that *no radioactive or disintegrating process can occur which, according
to Aston's measurements, can possibly produce
a radiation more than one-fourth as energetic,*
i. e., *as penetrating, as our softest observed cosmic rays nor more than one-thirtieth as energetic
as our most penetrating observed cosmic rays.*
On the other hand, the creation of the common
elements out of hydrogen should, according to
Aston's atomic mass measurements and Einstein's equation, release energies of just about
the observed penetrating power—quite accurately the observed penetrating power according
to the best available formulæ. But the exactness
of the fit is not at this stage very important. The
illuminating facts are, first, that only the *atom-
building process* can produce cosmic-ray *bands*
of anything like high enough energies, and,
second, that the *sequence* of observed cosmic
ray frequencies fits quite nicely the sequence of
the atomic weights of *the abundant* elements,

which are very few in number, for probably more than 95 per cent of all matter consists of oxygen, magnesium, silicon and iron. Our conclusion is that *the evidence is very strong that the cosmic rays are the subatomic radio signals broadcasting the continuous creation of these elements somewhere.*

But where? Apparently not in the sun or in the stars, since all observers agree that the sun, the great hot mass just off our bows, affects by his presence *not in the slightest* the stream of cosmic rays flowing into the earth. Further, Dr. Cameron and I can find no effect of the Milky Way, nor of the nearest extragalactic system, the spiral nebula of Andromeda. We think, therefore, that the high temperatures of the stars are inimical to the clustering processes required for the building of the common elements out of hydrogen, and we conclude that *the intensely cold regions of interstellar space probably furnish the essential conditions for such atom-building.*

Whether or not the foregoing conclusion is correct, Einstein's equation and Aston's curve alone, the former due partially to Mr. Michelson, the latter representing superrefinement in physical measurement of just the sort that Mr. Michelson is famous for, enable us to draw one definite and very important conclusion, namely, that *there is no energy available to man through the disintegration of any of the common elements.* Man will presumably some day learn to disintegrate the elements, but he will have to expend energy upon them to do it. *There is no appreciable energy available to man through atomic disintegration.* Radium, it is true, releases about a million times as much energy per gram in disintegrating as carbon does in burning, but there isn't enough of it, nor of any radioactive substance, to more than keep a few corner pop-corn men continuously going.

On the other hand, a practically unlimited supply of energy would be available to man if the hydrogen in water and elsewhere on the earth could unite, here on earth, to form helium, nitrogen, oxygen and the other common elements. This, we think, is just what is happening in interstellar space and thereby producing the observed cosmic rays, but the foregoing cosmic-ray facts seem to indicate *that this proc-*

ess cannot take place on earth, and if this is true, then man will, in the future as in the past, depend entirely upon the sun for his supply of available energy. To bring us up against such *realities* is the mission of men like Michelson. Inspiring realities they are, too, and their economic values are well-nigh unlimited, for we can direct our own efforts and our energies to better advantage with that knowledge. We have not yet begun to utilize the solar energy that is available to us, and we shall do it better with the knowledge that it is probably all we have.

Michelson's economic value! In the last analysis there is nothing that is practically important at all except our *ideas*, our group of concepts about the nature of the world and our place in it, for out of these springs all our conduct. There is not an idea that I have advanced to-night, a conclusion that I have drawn from Einstein's equation, from Aston's curve, from cosmic ray data, that would have been possible had not somebody driven to the limit the precision of physical measurement, and much of it became possible because of Michelson's own superrefined experiments—so true has it been proven to be that human progress "grows out of measurements made in the sixth place of decimals." Not he, nor anybody else, saw at the time what bearings the results would have. He merely felt in his bones, or knew in his soul, or *had faith to believe* that accurate knowledge was important. But some of the bearings have already appeared and others will continue to be found for ages yet to be.

I personally owe *everything* to the fact that thirty-two years ago Mr. Michelson took me into his nest at the University of Chicago, and I personally believe that the United States has not had in this generation a greater economic asset than Albert A. Michelson.

VIII

THREE GREAT ELEMENTS IN HUMAN PROGRESS

THERE are three ideas which seem to me to stand out above all others in the influence they have exerted and are destined to exert upon the development of the human race. They have appeared at widely separated epochs because they correspond to different stages in the growth of man's knowledge of himself and of his world. Each of these ideas can undoubtedly be traced back until its origins become lost in the dim mists of prehistoric times, for the sage and the prophet, the thinker and the dreamer, have probably existed since the days of the caveman, and the first has always seen, the second felt, truth to which his times were wholly unresponsive. But it is a general rule that *when the times are ripe* an idea that may have been adumbrated in individual minds millenniums earlier begins to work its way into the consciousness of the race as a whole, and from that time on exerts a powerful influence upon the springs of human progress. In this sense, these three ideas may be called discoveries and times may be set at which they begin to appear. The first of these, and the most important of the three, is the gift of religion to the race; the other two sprang from the womb of science. They are

1. The idea of the Golden Rule;
2. The idea of natural law;
3. The idea of age-long growth, or evolution.

The first idea, namely, that one's own happiness, one's own most permanent satisfactions are to be found through trying to forget oneself and seeking, instead, the common good, this altruistic ideal is so contrary to the immediate promptings of the animal within us that it is not strange that it found little place in the thinking or acting of the ancient world, nor for that matter, in spite of the professions of Christianity, in the acting of the modern world either. There will be common consent, however, that the greatest, most consistent, most influential proponent of this idea who has ever lived was Jesus of Nazareth. Buddha, Confucius, Socrates, all had now and then given voice to it, but Jesus made it the sum and substance of his whole philosophy of life. When he said, "All things whatsoever ye would that men should do unto you even so do ye also unto them, *for this is the law and the prophets*," I take it that he meant by that last phrase that this precept epitomized in his mind all that had been commanded and foretold—that it embodied the summation of duty and of aspiration.

Now when the life and teachings of Jesus became the basis of the religion of the whole western world, an event of stupendous importance for the destinies of mankind had certainly taken place, for a new set of ideals had been definitely and officially adopted by a very considerable fraction of the human race, a fraction which will be universally recognized to have held within it no small portion of the world's human energies and progressive capacities, and which has actually been to no small degree determinative of the direction of human progress.

The significance of this event is completely independent even of the historicity of Jesus. The service of the Christian religion, my own faith in essential Christianity would not be diminished one iota if it should in some way be discovered that no such individual as Jesus ever existed. If the ideas and ideals for which he stood sprang up spontaneously in the minds of men without the stimulus of a single great character, the result would be even more wonderful and more inspiring than it is now, for it would mean that the spirit of Jesus actually is more widely spread throughout the world than we realize. In making this statement, I am endeavoring to say just as positively and emphatically as I can, that the credentials of Jesus are found wholly in his teachings and in his character as recorded by his teachings, and not at all in any real or alleged historical events. And in making that affirmation, let me also emphasize the fact that I am only paraphrasing Jesus' own words when he refused to let his disciples rest his credentials upon a sign.

My conception then of the essentials of religion, at least of the Christian religion, and no other need here be considered, is that those essentials consist in just two things: first, in inspiring mankind with the Christlike, *i. e.*, the altruistic *ideal*, and that means specifically, *concern for the common good* as contrasted with one's own individual impulses and interests, wherever in one's own judgment the two come into conflict; and second, inspiring mankind to *do*, rather than merely to think about, its duty, the definition of duty for each individual being what he himself conceives to be for the common good. In three words, I conceive the essential task of religion to be "to develop the consciences, the ideals and the aspirations of mankind."

It is very important to notice that in the definitions I have given duty has nothing to do with what somebody else conceives to be for the common good, *i. e.*, with morality in the derivative sense of the mores of a people. Endless confusion and an appalling amount of futility gets into popular discussion merely because of a failure to differentiate between these two conceptions. As I shall use the words, then, moral and immoral, or moral right and wrong, are purely subjective terms. The question of what actually is for the common good is the whole stupendous problem of science in the broad sense of that term, *i. e.*, of knowledge, and has nothing to do with religion or with morals as I am using these words. There are only two kinds of immoral conduct. The first is due to indifference, thoughtlessness, failure to reflect upon what is for the common good, in other words, careless, impulsive, unreflective living on the part of people who know that they ought at least to try to think things through. I suspect that ninety-nine per cent of all immorality is of this type and that this furnishes the *chief reason for religious effort and the chief field for religious activity.* For both example and precept unquestionably have the power to increase the relatively small fraction of the population that attempts to be reflectively moral. The second type of immorality is represented by "the unpardonable sin" of which Jesus spoke, viz., deliberate refusal, after reflection, to follow the light when seen.

Thus far, I have been dealing only with what seem to me to be obvious facts, mere platitudes, if you will, for the sake of not being misunderstood when I speak about the essentials of religion. I am not at this moment concerned with

how far the *practice* of religion has at times
fallen short of the ideals stated in the foregoing
essentials. I am now merely reaffirming the
belief with which I began, and which I suspect
that, after the foregoing explanations, not many
will question, though I know there are some
who will, that the discovery of the foregoing
ideals, and their official adoption as the basis of
the religion of the Western World has within
the past two thousand years exercised a stupendous influence upon the destinies of the race.

But I am going to go farther and to express
some convictions about the relation of those
ideals not only to the past but, also, to the present and to the future. I am going to affirm
that those ideals are the most potent and significant element in the religion of the Western
World to-day. It is true that many individual
western religions contain some elements in ad-
dition to these, some of them good, some harm-
less, some bad, and that the good and the bad
are so mixed in some of them that it is not
always easy, even from my own point of view,
to determine whether a given branch of religion
is worthwhile or not. Nevertheless, looking at

western religion as a whole, the following facts seem to me obvious and very significant.

First, that if the basis of western religion is to be found in the element that is common to all its branches, then the one indispensable element in it now is just that element that formed the centre of Jesus' teaching, and that I have called above the essence of religion; second, that no man who believes in the fundamental value for the modern world of the essentials of religion as defined above, and in the necessity for the definite organization of religion for the sake of making it socially effective, needs to withdraw himself from the religious groups, and thereby to exert his personal influence against the spread of the essential religious ideals, since in America, at least, he will have no difficulty in finding religious groups who demand nothing of their adherents more than the belief in the foregoing ideals, coupled with an honest effort to live in conformity with them; third, that a very large fraction of the altruistic and humanitarian and forward-looking work of the world, in all its forms, has to-day its mainsprings in the Christian churches. My own judgment is that about ninety-five per cent of it has come, and is coming, directly or indirectly, from the influence of organized religion in the United States. My own judgment is that, if the influence of American churches in the furtherance of socially wholesome and forward-looking movements, in the spread of conscientious and unselfish living of all sorts, were to be eliminated, our democracy would in a few years become so corrupt that it could not endure. These last two are, however, merely individual judgments, the correctness of which I cannot prove. Some will, no doubt, differ with me in them.

Now looking to the influence of religion in the future, I have in the preceding found the essence of the gospel of Jesus in the Golden Rule, which, broadly interpreted, means the development in the individual of a sense of social responsibility. Civilization itself is dependent in the last analysis primarily upon just this thing. The change from the individual life of the animal to the group life of civilized man, especially in a world of science, a life of ever-expanding complexity as our scientific civilization advances, is obviously impossible unless, in general, the individual learns, in ever-increasing measure, to subordinate his impulses and interests to the furtherance of the group life. The reason that the western world, which has led, as we westerners think, in the development of civilization, adopted Christianity as its religion is to be found in the last analysis, I suspect, in the fact that western civilization with its highly organized group life found that it could not possibly develop without it; and if this is so, the future is certainly going to need the essentials of Christianity even more than the past has needed them. In other words, the job that the churches in the past have been in the main trying to do, and the job that, I think, in spite of their weaknesses and follies, they have in the main succeeded fairly well in doing; namely, the job of developing the consciences, the ideals, and the aspirations of mankind, must be done by some agency in the future even more effectively than it has been done in the past.

There are just two ways in which this can be done; namely, first, by destroying organized religion as Russia has recently been attempting to do, and building upon its ruins some other organization which will carry on the work that the church has in the main done in the past, some other organization which will embody the essentials of religion, but be free from its faults. The second way is to assist organized religion *as it now exists* to eliminate its faults and to be more effective in emphasizing and in spreading, with ever-increasing vigor, its essentials. The second method may perhaps be impossible in some countries. I should need to know those countries better than I do now before I could express an opinion. But, for our own country, I feel altogether sure of my ground, and I take it that most thinking men will agree with me that the second way is the only feasible way.

In the United States organized religion has already undergone an amazing evolution and has thus shown its capacity to evolve. It first sloughed off, or had cut away from it, the awful incubus of political power when the complete separation of church and state was decreed by the far-visioned men who made our Constitution. Second, it has to a considerable degree—

much more than in many countries—freed itself from the shackles that are imposed by central authority and vested rights and has thus
left itself free to evolve. Third, it has within
recent years been rapidly freeing itself, despite
some sporadic indications to the contrary, from
the curse of superstition, and getting nearer and
nearer to the essentials of religion. Finally, if
the growth of modern science has taught anything to religion and to the modern world, it
is that *the method of progress is the method of
evolution, not the method of revolution.* Let
every man reflect on these things well before he
assists by his influence in stabbing to death or
in allowing to starve to death, organized religion in the United States.

Thus far I have presented the most outstanding and conspicuous contribution of religion to
the development of the race. The two discoveries listed with that of the Golden Rule in my
opening sentence introduce us to the very mainsprings of the contribution of science to human
progress. The idea that God, or Nature, or the
Universe, whatever term you prefer, is not a
being of caprice and whim as had been the case

in all the main body of thinking of the ancient world, but is, instead, a God who rules through law, or a Nature capable of being depended upon, or a universe of consistency, of orderliness and of the beauty that goes with order—that idea has *made* modern science, and it is unquestionably the foundation of modern civilization.

It is because of this discovery, or because of the introduction of this idea into human thinking, and because of the *faith* of the scientist in it, that he has been able to harness the forces of nature and to make them do the work that enslaved human beings were forced to do in all preceding civilizations.

Yes, and much more than this, for it is not merely the material side of life that this idea has changed. It has also revolutionized the whole mode of thought of the race. It has changed the philosophic and religious conceptions of mankind. It has laid the foundations for a new and a stupendous advance in man's conception of God, for a sublimer view of the world and of man's place and destiny in it. The anthropomorphic God of the ancient world, the God of human passions, frailties, caprices, and whims is gone, and obviously with it the old duty, namely, merely or chiefly the duty to propitiate him, so that he may be induced to treat you, either in this world or the next, or in both, better than he treats your neighbor. Can any one question the advance that has been made in the diminishing prevalence of these mediæval, essentially childish, and essentially selfish ideas? The new God is the God of law and order, the new duty to know that order and to get into harmony with it, to learn how to make the world a better place for mankind to live in, not merely how to save your individual soul.[1] How-

[1]"Concerning what ultimately becomes of *the individual* in the process, science has added nothing and it has subtracted nothing. So far as science is concerned religion can treat that problem precisely as it has in the past, or it can treat it in some entirely new way if it wishes. For that problem is entirely outside the field of science now, though it need not necessarily always remain so. Science has undoubtedly been responsible for a certain change in religious thinking as to the relative values of individual and race salvation. For obviously by definitely introducing the most stimulating and inspiring motive for altruistic effort which has ever been introduced, namely, the motive arising from the conviction that we ourselves may be vital agents in the march of things, science has provided a reason for altruistic effort which is quite independent of the ultimate destination of the individual and is also much more alluring to some sorts of minds than that of singing hosannas forever around the throne. To that extent science is undoubtedly influencing and changing religion quite profoundly now. The emphasis upon making this world better is certainly the dominant and characteristic element in the religion of to-day."—ROBERT A. MILLIKAN, "Evolution in Science and Religion" (pp. 83 and 84), Yale University Press, 1927.

[179]

ever, once destroy our confidence in the principle of uniformity, our belief in the rule of law, and our effectiveness immediately disappears, our method ceases to be dependable, and our laboratories become deserted.

I am not worrying here over the recent introduction of the so-called "principle of uncertainty" in microscopic processes, an event that is causing so much excitement among physicists just now. This may indeed be consoling or, at least, illuminating to those non-physicists who have been worrying their heads over their inability to reconcile the principle of law with the facts of free-will and of responsibility. We physicists have had much worse contradictions than these to put up with in the subject of physics alone, as for example, the reconciliation of the wave theory of light with the essentially corpuscular light-quant theory. Experiment has told us that both theories are right, and we have had the limitations of our knowledge jolted into us enough times lately in physics to believe it, in spite of our inability to see as yet just how the reconciliation is to be made.

This fact worries Mr. Mencken, as it does all

essentially assertive (*i. e.*, dogmatic), minds, so that in a recent review of Eddington's extraordinarily profound book, "The Nature of the Physical Universe," he calls for another Huxley to tell us just exactly what is what in physics. But physicists have never been strong on dogmatism, not even in Huxley's day, and they are much less so now than then. We admit, to the complete bewilderment of minds like Mr. Mencken's, that we do not know everything yet.

In this book, Eddington points out that it may be illuminating to those who worry about free-will and determinism to know that in the subject of modern statistics, the behavior of a very large number of human beings such, for example, as the percentage of them that will get married per year is accurately predicted, though the behavior of a particular individual in the group is completely unpredictable and his choice unhampered. Here is certainly a specific illustration of the co-existence of the reign of law with the practical freedom of choice which each individual knows he has. But I don't think this particular problem ever worried the physicist, for he has always known *that his igno-rance was as yet quite ample enough to cover the links in the reconciliation that must exist.* Eighteenth-and-nineteenth century materialism never had any lure for him, for it always represented quite as pure dogmatism—assertiveness without knowledge—as did mediæval theology, *and modern developments have pushed it completely out of sight.* For *matter* is no longer a mere game of marbles played by blind men. An atom is now an amazingly complicated *organism*, possessing many interrelated parts and exhibiting many functions and properties— energy properties, radiating properties, wave properties, and other properties quite as mysterious as any that used to masquerade under the name of "mind," so that the phrases, "all is matter" and "all is mind" have now become merely shibboleths completely devoid of meaning.

However, whether the principle of determinism applies to infinitely minute, practically unattainable processes or not is not here important. For it is the existence of the *idea* of natural law or orderliness, with which we are concerned rather than with the proof of its universality, and no one who has any conception of what science has done since about 1600 A. D., the date at which this idea first began to spread throughout the consciousness of mankind, will be likely to question my initial statement that it is one of the three ideas which, whether true or false as a *universal* generalization, has at least exerted, and is undoubtedly still destined to exert, a stupendous influence upon the destinies of mankind.

The third, or evolutionary idea, is the youngest of the two great ideas born of modern science. It is not yet one hundred years old. Introduced by Darwin solely in its application to biological evolution, as discovery after discovery in modern science has pushed back farther and farther the age of the stars, the age of the solar system, the age of the earth, the age of the rocks, of fossil life, of prehistoric man, of recorded history, of social institutions, the evolutionary theory has come to dominate in a very broad way almost every aspect of human thought. We have come to the realization, not only that if biological forms, but also if social institutions like the family, the state, religion, or even war, have survived, it is because, after ages of trial in which many other institutions have competed with them and disappeared, they have had survival value. We have come to *study* institutions to see *why* they have survived. We have come to realize that if we wish to eliminate an old institution, like war, for example, we are not likely to succeed simply by wishing it gone, nor indeed simply by pacifistic propaganda of any sort. We are only likely to succeed if the conditions which gave it its survival value have been, or can be, eliminated. Hence, the establishment of a League of Nations, of a World Court, and the like, aimed precisely at eliminating some, at least, of these conditions. In my judgment, however, war is now in process of being abolished, chiefly by the relentless advance of modern science, *the principle diverter of man's energies and interests from the warlike to the peaceful arts*. War will disappear, like the dinosaur, when changes in world conditions, such as are now being brought about primarily by the growth of modern science and its applications—changes due to the advent of world-wide, nearly instantaneous intercommunication, and to the enormous modern stimulation of international trade and commerce, bringing with it a sense of interdependence and of the necessity of international understandings—when these changes have destroyed its survival value.

Again, because of the growth of this evolutionary idea in human thinking, we have come to see that an institution like religion, in so far as it deals with conceptions of God, the integrating factor in this Universe of atoms and of ether, and of mind, and of ideas, and of duties, and of intelligence, has not been and cannot be a fixed thing, that it has been continually changing with the growth of human knowledge, and that it will continue to expand as knowledge continues to grow.

I have thus presented the most outstanding and conspicuous contribution of religion to human progress, and the two most representative and significant contributions of science to human progress, and we are now ready to ask how these contributions are interrelated. The answer is altogether obvious. The world of science dominated by the reign of law has necessitated the increasing association of men into co-operating groups, but the effectiveness of those groups, indeed the whole group life, becomes at once impossible unless the altruistic ideals of religion, the sense of social responsibility, permeates the whole, while the evolutionary concept, the last contribution of science, is absolutely essential to an understanding of the development both of religion and of science. In a word, these three ideas and ideals interlock everywhere in a mutually helpful way. Not one of them can have a normal and effective existence without the others.

Whence, then, arises this strange idea, so often heard in popular discussions, of an incompatibility between *science* and *religion*? Here, again, I think the answer is clear. There is obviously no incompatibility between science and the *essentials of religion* as I have defined them. But individual religions, or branches of a religion often contain more than these essentials. Every movement which becomes popular and gains large numbers of adherents inevitably draws into itself men who are not actuated solely, nor even at all, by its ideals, but who are using it to further their own ends. Those ends may be very worthy ones, arising from the best of motives in minds of restricted understanding or limited intelligence, or they may be very unworthy ones, such as the desire for personal aggrandizement or political power. Every one knows that the history of Christianity is not at all free even from influences of the latter sort. The so-called War of the Reformation is usually described as a religious war, and the horrors of it are sometimes attributed to the influence of Christianity; but I think that most historians will agree with the statement that it was not primarily a religious war at all, although both sides undoubtedly worked overtime, as they always do, to try to prove that God was on their side. In other words, religion was its shibboleth, not its cause. It represented in the first instance simply the terrific struggle of a group of northern princes to free themselves from the yoke of a southern power which had used the machinery of a religious organization for cementing and perpetuating its control.

Again, the anticlerical parties to-day in

many countries that possess them represent in part, at least, the efforts of reformers to break the *political* power of groups that have seized it and hold it in the *name* of religion, when the real issues obviously have nothing whatever to do with religion. Still again, Voltaire in his attack on the church was not attacking religious ideals in the least. He did not even call himself an atheist. He was far too intelligent a man for that. Such presumptuous assumption of complete knowledge of the whole of what there is, or is not, in the universe I have never heard made by any men of real knowledge and understanding, for such men are always both humble and reverent. Fulness of knowledge always and necessarily means some understanding of the depths of our ignorance, and that always is conducive to both humility and reverence.

If you and I lived in some countries to-day I have no doubt that we should be in the anticlerical groups, but it would not be because we had lost confidence in the essentials of religion, but rather because we thought that these essentials had become so buried under excrescences of the kind I have been describing, introduced into the organization of religion by well-meaning but unintelligent men, or by designing men, that the net result was harmful rather than socially helpful.

But I have here been talking, not about religion and science, but rather about organized religion and politics—a pair that all of us will agree ought never to have been mated, and where they have been so mated ought to be divorced with the same celerity that characterizes proceedings at Reno. Fortunately, this problem does not exist for us in the United States. I have introduced the subject merely to show how the essentials of religion may, and sometimes do, become lost in the *organization* of religion. Present-day Buddhism is, I suppose, a more striking illustration of this than is anything that can be found among the many ramifications of Christianity.

But by the very same method described above in the discussion of politics and religion there has grown up another excrescence upon the essentials of religion which introduces us at once into the very heart of the alleged conflict between science and religion. This has come about not so much, like the marriage of politics with religion, because of the selfishness and ambition of men (real motives though often masking even in the minds of their possessors under softer names) as through the ignorance of men.

The amazing insight of Jesus is revealed by the fact that he kept himself so free from creedal statements, particularly statements that reflected the state of man's knowledge or ignorance of the universe that was characteristic of his times. A large part of his sayings seem to us now, in spite of the enormous increase in our knowledge of the universe that has taken place since his time, to be just as true to-day as they seemed to be then. The things that a man does not say often reveal the understanding and penetration of his mind even more than the things that he says. The fact that Jesus confined himself so largely to the statement of truths that still seem to us to have eternal value is what has made him a leader and teacher of such supreme influence throughout the centuries. But his followers, unlike him, have throughout the past two thousand years in many instances *loaded* their various branches of

[190]

his religion with creedal statements which are full of their own wofully human frailties. The difference is so enormous as to justify calling his statements, as the world has been wont to do, Godlike in comparison.

What are, in contrast, these man-made creeds? Admittedly they have been written by *men*, groups of men called together for the purpose, men so uninspired that very few of them have ever left any lasting memory of themselves behind. How many people now know of any name that was ever associated with any one of them? These men have often reflected in detail in their creeds the state of knowledge, or the state of ignorance, of the universe, or of God, whichever term you prefer, characteristic of their times. If some one wishes me to change that last implied definition of Deity so as to make it read, the unifying principle in the universe, I shall not object, for that there is a unity, an interrelatedness, a wholeness to it all, we ourselves being but parts of that whole, is attested by all experience, including, I should like to add, the amazing new scientific developments in the fields of ether physics, relativity, and wave-mechanics. That is only my prosaic paraphrase of the lines of Tennyson, the poet of science, when he says:

"The sun, the moon, the stars, the hills and the plains,
Are not these, O Soul, the vision of Him who reigns?
The ear of man cannot hear, and the eye of man cannot see;
But if we could see and hear this vision—were it not He?
Speak to Him, thou, for He hears and spirit with spirit shall meet.
Closer is He than breathing, and nearer than hands and feet."

Now with the conception of God changing continuously as man has grown in knowledge from the time when he pictured his God in the form of a calf, or a crocodile, or a monstrous man, to the time when the poet described God, as above, as the soul of the universe, what must be the relation between science, or the ever-expanding knowledge of man, and the long since vanished conceptions of the universe, or of God, frozen in ancient man-made creeds? Obviously one of inescapable conflict. And in so far as these creedal excrescences have covered up, or displaced, the essentials of religion, there are obviously no alternatives except (1) to remove that sort of a deadening growth from the heart of religion, or (2) failing that possibility to desert a hopeless religion, or (3) to give up science.

A choice between the last two alternatives might in some countries be a necessity. But fortunately in the United States, which, being the widest flung democracy in the world, needs, and indeed must have, if it is to endure, the essentials of religion more than any other country, there is no such choice necessary. For in this country religion has been able to develop wholly untrammelled by political interference, and it has in many of its branches been absolutely free to develop without the restraining influence of central authority. I have myself belonged to two churches, one a Union church and one a Congregational church, both of which were unhampered by a creed of any sort. Other churches are continually revising or modifying their creeds with our growing knowledge.

Within the United States there is then not the slightest reason why religion cannot keep completely in step with the demands of our continuously growing understanding of the world. Here religious groups are to be found which correspond to practically every stage of the development of our knowledge and understanding. Here there is no need, in the case of any individual, of a clash ever arising between science and religion. Personally I believe that essential religion is one of the world's supremest needs, and I believe that one of the greatest contributions that the United States ever can, or ever will, make to world progress—greater by far than any contribution which we ever have made or ever can make to the science of government—will consist in *furnishing an example to the world of how the religious life of a nation can evolve intelligently, wholesomely, inspiringly, reverently, completely divorced from all unreason, all superstition, and all unwholesome emotionalism.*